NATURAL PRODUCTS PHARMACOLOGY AND PHYTOCHEMICALS FOR HEALTH CARE

Methods and Principles in Medicinal Chemistry

NATURAL PRODUCTS PHARMACOLOGY AND PHYTOCHEMICALS FOR HEALTH CARE

Methods and Principles in Medicinal Chemistry

Edited by

Debarshi Kar Mahapatra, PhD

Cristóbal Noé Aguilar, PhD

A. K. Haghi, PhD

AAP | APPLE ACADEMIC PRESS

First edition published 2021

Apple Academic Press Inc.
1265 Goldenrod Circle, NE,
Palm Bay, FL 32905 USA

4164 Lakeshore Road, Burlington,
ON, L7L 1A4 Canada

CRC Press
6000 Broken Sound Parkway NW,
Suite 300, Boca Raton, FL 33487-2742 USA

2 Park Square, Milton Park,
Abingdon, Oxon, OX14 4RN UK

Library and Archives Canada Cataloguing in Publication

Title: Natural products pharmacology and phytochemicals for health care : methods and principles in medicinal chemistry / edited by Debarshi Kar Mahapatra, PhD, Cristóbal Noé Aguilar, PhD, A.K. Haghi, PhD.

Names: Mahapatra, Debarshi Kar, editor. | Aguilar, Cristóbal Noé, editor. | Haghi, A. K., editor.

Description: Includes bibliographical references and index.

Identifiers: Canadiana (print) 20200363433 | Canadiana (ebook) 20200363506 | ISBN 9781771889018 (hardcover) | ISBN 9781003057932 (ebook)

Subjects: LCSH: Biological products. | LCSH: Drug development. | LCSH: Pharmaceutical chemistry.

Classification: LCC RM301.25 .N38 2021 | DDC 615.1/9—dc23

Library of Congress Cataloging-in-Publication Data

Names: Mahapatra, Debarshi Kar, editor. | Aguilar, Cristóbal Noé, editor. | Haghi, A. K., editor.

Title: Natural products pharmacology and phytochemicals for health care : methods and principles in medicinal chemistry / edited by Debarshi Kar Mahapatra, Cristóbal Noé Aguilar, A.K. Haghi.

Description: First edition. | Palm Bay, FL : Apple Academic Press, 2021. | Includes bibliographical references and index. | Summary: "Medicinal chemistry and pharmacology are closely associated fields, and the use of natural products for their medicinal properties is ever-growing. The study of drugs from natural products and their effects on the living body are explored in this volume. The book looks into the research, discovery and characterization of the chemicals that exhibit biological effects. Providing an informative compilation of research, valuable case studies, and review of existing literature in the area, the book focuses on the ethnobotanical uses of natural products and phytochemicals for health care, including applications for diabeties, ulcers, wound healing, managing chronic alcoholism, hemorrhoidal treatment, cancer mitigation, pain management, immunotherapy, and more. The book briefly describes bioinformatics, artificial intelligence, machine learning, innovations, and societal applications. Natural Products Pharmacology and Phytochemicals for Health Care: Methods and Principles in Medicinal Chemistry provides a practical and comprehensive overview of the daily issues facing pharmaceutical researchers and chemists. This volume provides new coverage of some of the latest technologies and approaches in drug discovery"-- Provided by publisher.

Identifiers: LCCN 2020046631 (print) | LCCN 2020046632 (ebook) | ISBN 9781771889018 (hardcover) | ISBN 9781003057932 (ebook)

Subjects: MESH: Biological Products | Drug Discovery

Classification: LCC RM301.25 (print) | LCC RM301.25 (ebook) | NLM QV 241 | DDC 615.1/9--dc23

LC record available at https://lccn.loc.gov/2020046631

LC ebook record available at https://lccn.loc.gov/2020046632

ISBN: 978-1-77188-901-8 (hbk)
ISBN: 978-1-00305-793-2 (ebk)

About the Editors

Debarshi Kar Mahapatra, PhD
Assistant Professor, Department of Pharmaceutical Chemistry, Dadasaheb Balpande College of Pharmacy, Rashtrasant Tukadoji Maharaj Nagpur University, Nagpur, Maharashtra, India

Debarshi Kar Mahapatra, PhD, is currently an Assistant Professor in the Department of Pharmaceutical Chemistry at Dadasaheb Balpande College of Pharmacy, Rashtrasant Tukadoji Maharaj Nagpur University, Nagpur, Maharashtra, India. He was formerly an Assistant Professor in the Department of Pharmaceutical Chemistry, Kamla Nehru College of Pharmacy, RTM Nagpur University, Nagpur, India. He has taught medicinal and computational chemistry at both the undergraduate and postgraduate levels and has mentored students in their various research projects. His area of interest includes computer-assisted rational designing and synthesis of low molecular weight ligands against druggable targets, drug delivery systems, and optimization of unconventional formulations. He has published research, book chapters, reviews, and case studies in various reputed journals and has presented his work at several international platforms, for which he has received several awards by a number of bodies. He has also authored the book titled *Drug Design*. Presently, he is serving as a reviewer and editorial board member for several journals of international repute. He is a member of a number of professional and scientific societies, such as the International Society for Infectious Diseases (ISID), the International Science Congress Association (ISCA), and ISEI.

Cristóbal Noé Aguilar, PhD
Director of Research and Postgraduate Programs at Universidad Autonoma de Coahuila, Mexico

Cristóbal Noé Aguilar, PhD, is a Director of Research and Postgraduate Programs at Universidad Autonoma de Coahuila, Mexico. Dr. Aguilar has published more than 160 papers in indexed journals, more than 40 articles in Mexican journals, and 250 contributions in scientific meetings. He has also published many book chapters, several Mexican books, four editions of international books, and more. Prof. Aguilar is a member of level III of S.N.I. (National System of Researchers of Mexico). He has been awarded several prizes and awards; the most important are the National Prize of Research 2010 from the Mexican Academy of Sciences; the Prize "Carlos Casas Campillo 2008" from the Mexican Society of Biotechnology and Bioengineering; National Prize AgroBio–2005 and the Mexican Prize in Food Science and Technology. Dr. Aguilar is a member of the Mexican Academy of Science, the International Bioprocessing Association, Mexican Academy of Sciences, Mexican Society for Biotechnology and Bioengineering, and the Mexican Association for Food Science and Biotechnology. He has developed more than 21 research projects, including six international exchange projects. He has been an advisor for PhD, MSc, and BSc theses. Dr. Aguilar earned his chemist designation from the School of Chemistry at the Universidad Autónoma de Coahuila, México. He attended the MSc Program in Food Science and Biotechnology that was held at the Universidad Autónoma de Chihuahua, México. His PhD in Fermentation Biotechnology was awarded by the Universidad Autónoma Metropolitana, Mexico.

A. K. Haghi, PhD

Professor Emeritus of Engineering Sciences, Former Editor-in-Chief, International Journal of Chemoinformatics and Chemical Engineering and Polymers Research Journal; Member, Canadian Research and Development Center of Sciences and Culture

A. K. Haghi, PhD, is the author and editor of over 200 books, as well as over 1000 published papers in various journals and conference proceedings. Dr. Haghi has received several grants, consulted for a number of major corporations, and is a frequent speaker to national and international audiences. Since 1983, he served as a professor at several universities. He is former Editor-in-Chief of the *International Journal of Chemoinformatics and Chemical Engineering* and *Polymers Research Journal* and is on the editorial boards of many international journals. He is also a member of the Canadian Research and Development Center of Sciences and Cultures (CRDCSC), Montreal, Quebec, Canada.

Contents

Contributors

Mojabir Hussen Ansari
Gurunanak College of Pharmacy, Dixit Nagar, Nagpur, 440026, Maharashtra, India

Vivek Asati
Department of Pharmaceutical Chemistry, NRI Institute of Pharmacy, Bhopal, Madhya Pradesh, 462021, India

Sanjay Kumar Bharti
Institute of Pharmaceutical Sciences, Guru Ghasidas Vishwavidyalaya (A Central University), Bilaspur, Chhattisgarh, 495009, India

Shilpa S. Borkar
Department of Pharmacology, Kamla Nehru College of Pharmacy, Nagpur – 441108, Maharashtra, India

Gloria Castellano
Departamento de Ciencias Experimentales y Matemáticas, Facultad de Veterinaria y Ciencias Experimentales, Universidad Católica de Valencia San Vicente Mártir, Guillem de Castro-94, E-46001 València, Spain

Zanariah Hashim
Food and Biomaterial Engineering Research Group, School of Chemical and Energy Engineering, Faculty of Engineering, Universiti Teknologi Malaysia, 81310 Johor Bahru, Malaysia

Sameer A. Hedaoo
Department of Pharmaceutical Chemistry, Gurunanak College of Pharmacy and Technical Institute, Nagpur – 440026, Maharashtra, India

Yanti Maslina Mohd Jusoh
Food and Biomaterial Engineering Research Group, School of Chemical and Energy Engineering, Faculty of Engineering, Universiti Teknologi Malaysia, 81310 Johor Bahru, Malaysia

Paras Kothari
Department of Pharmaceutics, Gurunanak College of Pharmacy and Technical Institute, Nagpur – 440026, Maharashtra, India

Debarshi Kar Mahapatra
Assistant Professor, Department of Pharmaceutical Chemistry, Dadasaheb Balpande College of Pharmacy, Nagpur, 440037, Maharashtra, India, E-mail: mahapatradebarshi@gmail.com

Ida Idayu Muhamad
Food and Biomaterial Engineering Research Group, School of Chemical and Energy Engineering, Faculty of Engineering, Universiti Teknologi Malaysia, 81310 Johor Bahru, Malaysia

Eraricar Salleh
Food and Biomaterial Engineering Research Group, School of Chemical and Energy Engineering, Faculty of Engineering, Universiti Teknologi Malaysia, 81310 Johor Bahru, Malaysia

Ankita Soni
Department of Pharmaceutics, Gurunanak College of Pharmacy and Technical Institute, Nagpur – 440026, Maharashtra, India

Francisco Torrens
Institut Universitari de Ciència Molecular, Universitat de València, Edifici d'Instituts de Paterna, P.O. Box 22085, E-46071 València, Spain, E-mail: torrens@uv.es

Seema Usare
Department of Pharmacology, Kamla Nehru College of Pharmacy, Nagpur – 441108, Maharashtra, India

Dayang Norulfairuz Abang Zaidel
Food and Biomaterial Engineering Research Group, School of Chemical and Energy Engineering, Faculty of Engineering, Universiti Teknologi Malaysia, 81310 Johor Bahru, Malaysia, Tel: +607 5535503; E-mail: dnorulfairuz@utm.my

Abbreviations

ACh	acetylcholine
AD	Alzheimer's
ADP	adenosine diphosphate
AEs	adverse effects
AI	artificial intelligence
AIA	anti-inflammatory activity
AIDS	acquired immunodeficiency syndrome
AMA	anti-melanogenic action
ANOVA	analysis of variance
AOA	AO activity
AOs	antioxidants
ART	artemisinin
ATM	African TM
ATs	analytical techniques
BA	bioactive
BA	brown algae
BC	breast cancer
BD	big data
BD	biodiversity
BDA	BD analytics
BDMC	bisdemethoxycurcumin
BE	bioeconomy
BI	bioinformatics
BPA	bisphenol A
BR	bioreceptor
BSs	biosensors
CAM	complementary and alternative medicine
CBE	circular bioeconomy
CC	cancer of colon
CD	compact disc
CDKs	cyclin-dependent-kinases
CE	circular economy
CeC	cervical cancer
CH_2O	carbohydrate
CHOL	cholesterol

CMC	carboxymethyl cellulose
CNS	central nervous system
CP	chronic pain
CRC	colorectal cancer
CRISPR	clustered, regularly interspaced, short palindromic repeat
Cs	conclusions
CT	chemotherapy
CVD	chemical vapor deposition
CVDs	cardiovascular diseases
DC	digestive cancer
DES	diethylstilbestrol
DHA	docosahexaenoic acid
DHM	dihydromyricetin
DM	dark matter
DM	diabetes mellitus
DMC	demethoxycurcumin
dMMR	deficient mismatch repair
DPPH	2,2-diphenyl-1-picrylhydrazyl
DySCoVir	dynamical systems and computational virology
EB	ethnobotany
EC	endometrial cancer
EC	epicatechin
ECHA	European Chemicals Agency
EGC	epigallocatechin
EGCG	epigallocatechin gallate
EMFs	electromagnetic fields
EMR	electromagnetic radiation
EOs	essential oils
EP	ethnopharmacology
ESI	electrospray ionization
ETs	electronic tongues
FFs	force fields
FHOs	fractal hybrid orbitals
FLU	fluorescence
FOS	fructooligosaccharides
FRs	free radicals
FuFoSE	functional food science in Europe
GA	geriatric assessment
GA	green alga

GABA	γ-aminobutyric acid
GC	gallocatechin
GC/MS	gas chromatography/mass spectrometric
GDP	gross domestic product
GOx	glucose oxidase
GPR	G-protein coupled receptor
GR	graphene
GRs	graphemes
GSIS	glucose-stimulated insulin secretion
GUC	genitourinary cancer
H	hemagglutinin
H	hypotheses
H1N1	swine-origin influenza A
HC	healthcare
HC	hydrocarbon
HHP	high hydrostatic pressure
HIV	human immunodeficiency virus
HIV-PR	HIV-1 protease
HMB	2-hydroxy-3-methyl-3-butenyl alkyl
HNC	head and neck cancer
HPLC	high-performance liquid chromatography
HPP	high-pressure processing
HR	high-risk
HRD	homologous recombination deficiency
HRT	hormone replacement therapy
HTST	high temperature short time
I^2SysBio-CRM	Integrative Systems Biology–Centre de Recerca Matemàtica
IBD	inflammable bowel disease
IBS	irritable bowel syndrome
ICIQ	Institut Català d'Investigació Química
ICT	information and communication technology
ILSI	International Life Science Institute
IS	immune system
ISs	integrated systems
IT	immunotherapy
JAR	just about right
LAM	labeled affective magnitude
LB	liquid biopsy
LC	lung cancer

LE	linear economy
LGTB	lesbian-gay-bisexual-transgender
LH	gonadotrophic hormone
LN	lymph node
LPS	level in lipo-PS
LTLT	low temperature long time
mBRCA	BRCA mutation
MC	miscellaneous cancer
MC1R	melanocortin-1 receptor
MDP	Mediterranean diet pyramid
MF	microfiltration
MHG	microwave hydro-diffusion and gravity
MIP	molecularly imprinted polymer
ML	machine learning
ML	monolayer
MMPs	matrix metalloproteinases
MOFs	metal-organic frameworks
MOR	μ-opioid receptor
MPs	medicinal plants
MRD	minimal residual disease
mRNA	messenger ribonucleic acid
MS	mass spectrometry
MSI	microsatellite instability
MTL	monoterpene lactone
MTT	3-(4,5-dimethylthiazol-2-yl)-2,5-diphenyltetrazolium bromide
N	neuraminidase
NA	noradrenaline
NAM	nicotinamide
NAs	nucleic acids
NCs	natural compounds
NDEA	N-nitrosodiethylamine
NF-κB	nuclear factor κ-light-chain-enhancer of activated B cells
Nivo/Ipi	nivolumab + ipilimumab
NOS	Nitric oxide synthases
NP	natural product
NRF2	nuclear factor (erythroid-derived 2)-like 2
NSP	non-starch polysaccharides
OC	ovarian cancer

OEST	estrogen
OS	overall survival
OS	oxidative stability
OS	oxidative stress
P3S	proteins' tertiary structure
PA	protective action
PAN	peroxyacetyl nitrate
PARP	poly [adenosine diphosphate (ADP)-ribose] polymerase
PC	prostate cancer
PCA	phytochemical analysis
PCA	principal component analysis
PCBs	polychlorinated biphenyls
PD	Parkinson's disease
PDL	programmed death-ligand
PF	pelvic floor
PFS	progression-free survival
Pj-EE	*Prasiola japonica* ethanol extract
PM	precision medicine
PME	pectin methylesterase
POD	peroxidase
PPM	personalized PM
PPO	polyphenol oxidase
PSs	polysaccharides
PTE	periodic table of the elements
PUFAs	polyunsaturated fatty acids
QDA	qualitative descriptive analysis
RF	radio frequency
RO	reverse osmosis
ROSs	reactive oxygen species
RoT	round table
RSF	relapse-free survival
RT	radiotherapy
RVC	relative volume change
SARs	structure-activity relationships
SDBC	Sunny Delight Beverages Co.
SDFs	soluble dietary fibers
SDs	synthetic drugs
SOCS	suppressor of cytokine signaling
SPR	surface plasmon resonance
STEM	science, technology, engineering, and mathematics

STL	sesquiterpene lactone
STS	science, technology, and society
SWNCs	single-wall carbon nanocones
SWNHs	single-wall carbon nanohorns
SWNTs	single-wall carbon nanotubes
SZT	streptozotocin
TCM	Chinese TM
TFC	total flavonoid content
TM	traditional medicine
TMV	tobacco mosaic virus
TNBC	triple-negative BC
TPC	total phenolic content
TSS	total suspended solids
UF	ultrafiltration
UTFs	ultra-thin films
UV-VIS	ultraviolet-visible
VdW	van der Waals
Vit	vitamin
VOs	vegetable oils

Preface

This book is a vital tool for all researching and studying pharmaceutical science and medicinal chemistry. It provides a comprehensive, balanced introduction to this evolving and multidisciplinary area of research.

Medicinal chemistry and pharmacology are closely associated fields. The study of drugs and their effects on the living body are explored in pharmacology. It involves the research, discovery, and characterization of the chemicals that exhibit a biological effect.

This volume provides a practical and comprehensive overview of the daily issues facing pharmaceutical researchers and chemists. In addition to its thorough treatment of basic medicinal chemistry principles, this volume provides new and expanded coverage of the latest technologies and approaches in drug discovery.

Providing a unique compilation of data, insightful case studies, and a review of existing literature in the area, the book will promote innovation in medicinal and pharmaceutical chemistry research, exploring the limitations of existing drugs and their improvement.

This book details several important medicinal plants, their occurrence, plant compounds, and their chemical structures, and pharmacological properties against various human diseases. This book is an indispensable tool for researchers, as well as graduates in various disciplines, including pharmacy, pharmacology, medicine, biotechnology, nutrition, and drug development.

In the first chapter, a short overview of anti-diabetic natural products is presented. It is shown that the natural medicinal product possesses many pharmacological activities like anti-diabetic, anti-ulcer, anti-tumor, anti-pyretic, anti-lipidemic, wound healing, etc. The natural products are widely used from therapeutic purpose from ancient time. Many of the natural products have a well-known mechanism for their activity but still, most of the natural products have not established mechanisms as well as an active therapeutic component for their activity, such type of natural products are mainly incorporated into work to establish their bioactive component by using a sophisticated instrument such as ultra-violet spectroscopy, mass spectroscopy, NMR, etc. Natural products also contain some wonderful drugs such as Tulsi, Neem, and Aloe vera, which possess great potential to therapeutic uses from the ancient times of mankind but still the mechanism some pharmacological activity of these plants is not known. In the future, a controlled study is required to prove the effectiveness of natural products under various conditions.

In Chapter 2, it was found that alcoholism is a serious problem throughout the world. It hampers the balance of the body and creates a dependency on it. The above information has included the botanical characteristics of the plant, which helps in the identification of the plant, ethnobotany which gives traditional use of the plant, and the reported activities of the plant. Various synthetic drugs are available for the treatment of alcoholism but they were severe side effects and contraindication regarding the use of these drugs. Nowadays the world is getting attention to natural remedies. The therapeutic effects of these natural remedies may be less pronounced at times than synthetic drugs, but the probability of adverse symptoms to a minimum. By this information, it can be concluded that in the core of nature there are so many plants, which possess anti-alcoholic property. Herbal medications are free from side effects and toxicity, unlike the synthetic ones. So these natural actives can be easily and safely replaced as anti-alcoholics as a part of treatment.

The herbs can be an effective aphrodisiac, moreover, isolation and identification of active constituents from plants may bring a dynamic change in the modern world. In Chapter 3, we have highlighted that *Lepidium meyenii* (usually called Maca) may be a Peruvian plant that grows high within the mountain chain, in areas over four, 4000 meters on top of water level. It contains most of the essential amino acids and twenty of the essential fatty acids, further as several vitamins, minerals, and several other sterols and glucosinolates required by man. The clinical studies of varied crude extracts from differing kinds of maca have multiple applications such as nutritionary, chemoprevention, aphrodisiac, and fertility-enhancing functions. The particular and distinctive unsaturated acids and amides found solely during this plant are the macaenes and macamides contained at intervals the hypocotyls.

New insights in the matter, plasma, quasi-two-dimensional Au film, drug product quality, and circular economy are presented in Chapter 4.

Hemorrhoids, also known as piles, are vascular structures in the anal canal which help with stool control. They become a disease when swollen or inflamed. Generally, there are two types of hemorrhoids and the sign and symptoms depend on the type present. Internal hemorrhoids are painless, often result in rectal bleeding when defecating. External hemorrhoids result in pain and swelling in the anal area. Male and female are both equally affected most often between 45 and 65 years of age, the condition lasts for a few days. A number of factors that lead to the diseased condition include constipation or diarrhea, lack of exercise, low fiber diets, increased intra-abdominal pressure, genetics, aging. Management of pain, inflammation, bleeding is a goal of treatment in hemorrhoids. Chapter 5 gives information about the pathophysiology and detailed review of treatment using herbal formulation and reveals their different pharmacological activities such as anti-inflammatory, analgesic, anti-oxidant, anti-microbial, immunosuppressant, and laxative properties.

In Chapter 6, a new development in precision, immunotherapy, immuno-oncology, and research in cancer are reported in detail. Oxidative stress-protective/anti-melanogenic effects of loliolide are investigated in Chapter 7.

In recent years, an extensive outburst of abundant infectious ailments across the world has created chaos between the populations. Chiefly, the residents in the tropical and sub-tropical areas are concerned by these virulent pathogens. In spite of the development of significant elementary acquaintance regarding a variety of pathogens, the existing chemotherapy is still unacceptable as a result of imperfect effectiveness, long-term action, high cost, and unwanted side-effects. Natural products have been found to express numerous pharmacological activities. Several alcoholic, hydroalcoholic, and aqueous extracts along with phytoconstituents are reported to exhibit anti-infective effects by modulating the number of pharmacotherapeutic targets. Chapter 8 focuses on the therapeutic perspectives of chalcone based natural compounds in exhibiting antiviral activity against rhinovirus, human immunodeficiency virus (HIV), influenza virus, poliovirus, dengue virus, and tobacco mosaic virus (TMV) by modulating vital therapeutic targets like HIV-1 protease, DEN-1 NS3 protease, and neuraminidase. These natural compounds serve as interesting leads for synthetic drug fabrication by various drug design, drug discovery, and drug development approaches as well as structure-activity relationships (SARs) predictions. These compounds from nature may serve as templates for further exploration of molecular targets, structural biology, mechanism of actions (MOAs), etc.

Ethnoscience, alternative medicine, inflammation, and pain models are discussed in Chapter 9. In Chapter 10, it is shown that plants form the basis of African traditional medicine, which contributed significantly to the reduction of mortality, morbidity, and disability owing to many diseases. Medicinal plants were reported to be an essential aspect of oral health delivery cutting across all facets of oral healthcare. One should avoid the medical practice that carries any supplementary description, since he will surely be paying expensive for something that no one can guarantee people to work. Among the therapies are well-known ones (homeopathy, acupuncture), and others more strange that somebody invents and applies them. Though traditional healers are trained in handling plant material, additional training is required for plants to be safe for medicine. Medicinal plants present benefits in the management of oral diseases but face challenges. Research on medicinal plants used in African traditional medicine in oral health is scanty. Mediterranean Diet Pyramid is the basis for integrative medicine for veterans with disabilities, but patient-centered and interprofessional approaches represent the real added value for the best healthcare management.

Recent reports have recognized that functional foods are the most emerging interest in food sectors. In particular, the functional beverage market is one of the fastest-growing segments within this sector. Beverages are categorized as the most dynamic functional food due to their convenience and the high possibility to meet market needs. Functional beverages have great potential to be incorporated with desirable nutritional ingredients and bioactive compounds such as vitamins, minerals, antioxidants, fatty acids, fiber, prebiotics, and probiotics. Chapter 11 highlighted the recent innovation and development of functional and medicinal beverages including the current processes and methods of developing successful functional and medicinal beverages. The topics include the definition of functional and medicinal beverages and the type of functional beverages available in the market, the market trends and overview, emerging trends in processing technology, quality, and safety issues in functional beverages prior to commercialization of products. It has become a challenge to maintain a good quality of functional and medicinal beverages especially in terms of its stability, shelf life, functionality, and safety. These issues must be taken seriously for improving the quality and health benefits of the beverages, as well as for the future of the food industry.

Estradiol, aging theory, women/LGBT difficulties, and motivation are discussed in Chapter 12. And finally, in the last chapter of this book, women, men, health, big data, taste to classify, and innovation are presented.

CHAPTER 1

A Short Overview on Anti-Diabetic Natural Products: Reviewing the Herbotherapeutic Potentials

MOJABIR HUSSEN ANSARI[1] and DEBARSHI KAR MAHAPATRA[2]

[1]*Gurunanak College of Pharmacy, Dixit Nagar, Nagpur, 440026, Maharashtra, India*

[2]*Assistant Professor, Department of Pharmaceutical Chemistry, Dadasaheb Balpande College of Pharmacy, Nagpur, 440037, Maharashtra, India, E-mail: mahapatradebarshi@gmail.com*

ABSTRACT

Plants contain various types of bioactive components that show many pharmacological activities. It provides instant people need and are commonly available and inexpensive. Also, medicinal plants have less side effects compared to synthetic organic chemicals. In the past decade, there has been tremendous growth in the implementation of natural products as a health regimen in Asia and European countries. Herbal natural products have a strong conceptual base and admire to be used as a drug in terms of safety and benefit for treating various diseases. Herbal medicines are becoming more popular due to its safety and efficacy during treatment. Although it contains various active chemical constituents which is useful in the management and treatment of most of diseases. Natural product mainly consists of flavonoids, tannins, saponins, fatty acids, ellagic acid, chebulic acid, alkaloids, glycosides, reserpine, vincristine, vinblastine, quinidine, etc. Different plants consist of different chemical constituent hence it gives different activities such as anti-ulcer, anti-microbial, anti-hypertensive, anti-diabetic, anti-cancer, anti-malarial, etc. Diabetes Mellitus is the largest chronic metabolic disorder affecting about 390 million populations across the globe and it is estimated to affect 592 million people by the year 2030. Many natural and semi-synthetic chalcones are important candidates that interfere with a therapeutic target such as PPAR-γ, DPP-4, α-glucosidase, PTP-1B, aldose reductase, enhances insulin production, and increase tissue sensitivity. Many organic molecules are available for the management and treatment of various types of diseases, it produces a more adverse effect than herbal medicine. Therefore, herbal medicines are more popular. Hence, major countries are contributed in the development of natural medicines.

1.1 INTRODUCTION

Natural products are the part of the human living system since the time origin. Plants give numerous products for human needs, most of them good and beneficial. Everything such as food, shelter, clothes, medicines, fuels, etc. comes out from plants directly and indirectly. Plant are the milestones of all living entities on this planet and an important resource for all living organisms as well as humans [1]. Medicinal plants as a part of traditional remedies possess the key role of human health care. Neanderthal made medicine by using medicinal plant and it is shown in molecular evidence. Modern human also implies medicinal plant and their part for the treatment of disease in the form of herbal medicine. It is not possible to pinpoint the percentage of traditional medicine use around the globe, evidence showed that the majority of the population in developing nations consistently based upon natural medicine and 50% of population rely on natural products in developed countries. Most of organic molecules has been developed from the plant part. From 1981 to 2002, 45% of drugs have been approved for infectious diseases that are obtained from the natural product as well as tropical forests contributed as 25% for the development of medicine [2]. A phenolic compound such as phenolic acid flavonoids, tannins, stilbenes, coumarins, lignans, and lignins are mainly present in part of natural products that show multiple biological effects [3]. In our routine nutraceutical diet like turmeric, cardamom, garlic, onion, ginger, tulsi, cloves, etc. contain medicinal value which is historically used for treatment of various diseases [4].

Diabetes Mellitus is a serious chronic metabolic disorder of fat, protein, and carbohydrates metabolism. It is characterized by an excess of blood glucose, because pancreatic β-cells does not produce sufficient insulin that is required to convert sugar, starch and other food into energy [5]. The recent review elaborate the use of natural products for therapeutics of diabetes. Flowers, fruits, roots of endemic plant species are the ethnomedicine which are introduced by local community in the form of crude extract for cure of diabetes [6]. Moreover, in some situation, plants extract itself treating diabetes-related disorders like polydipsia, polyuria, glucosuria, etc. *Charak samhita* reported the implementation of plant, herbs, and their various derivatives for the management of diabetes mellitus. About 450 plants have been implemented and nearly 750 recipes are presently used to manage diabetes mellitus of two-third populations in the globe. Leguminosae, Liliaceae, Cucurbitaceae, Lamiaceae, Moraceae, Asteraceae, Rosaceae, Euphorbiaceae, and Araliaceae are the plants family which have reported anti-diabetic activity [7]. *S. cumini, A. indica, O. tenuflorum, M. koenigii, L. usitatissimum*, etc. are having anti-diabetic activity, are already reported in Ayurveda [8].

1.2 MECHANISM OF ANTI-DIABETIC ACTIVITY OF SOME MEDICINAL PLANT

1.2.1 *GINSENG*

This medicinal plant is mostly used in research for anti-diabetic activity and it was found that the efficacy as well as the potency of the plant depends upon geographical location, dosage, and the type of diabetes. Ginseng saponins such as glycans, panaxans I, J, K, and L had a hypoglycemic activity (Figure 1.1). The ginsenoside Rh2 can stimulate the production of acetylcholine (ACh) from the nerve terminal. Released acetylcholine stimulates muscarinic (M3) receptor in β-cells of pancreas, therefore increase of excess insulin take place [9–11].

1.2.2 MOMORDICA CHARANTIA

Momordica charantia, it has long list of common name such as bitter guard, bitter melon, balsam pear or 'fugkwa.' The hypoglycemic activity is achieved due to vicine, polypeptide-P, and charantin containing mixed sterols that facilitate glucose uptake and glycogen formation in muscle cell and liver and reduce glucose synthesis (Figure 1.2). Bitter melon reduces the blood glucose concentration via suppression of STZ induces peroxidation and apoptosis in pancreatic β-cells, by increasing the glycogen formation in liver with activating hepatic glucokinase, hexokinase, and phospho-fructokinase [9, 12].

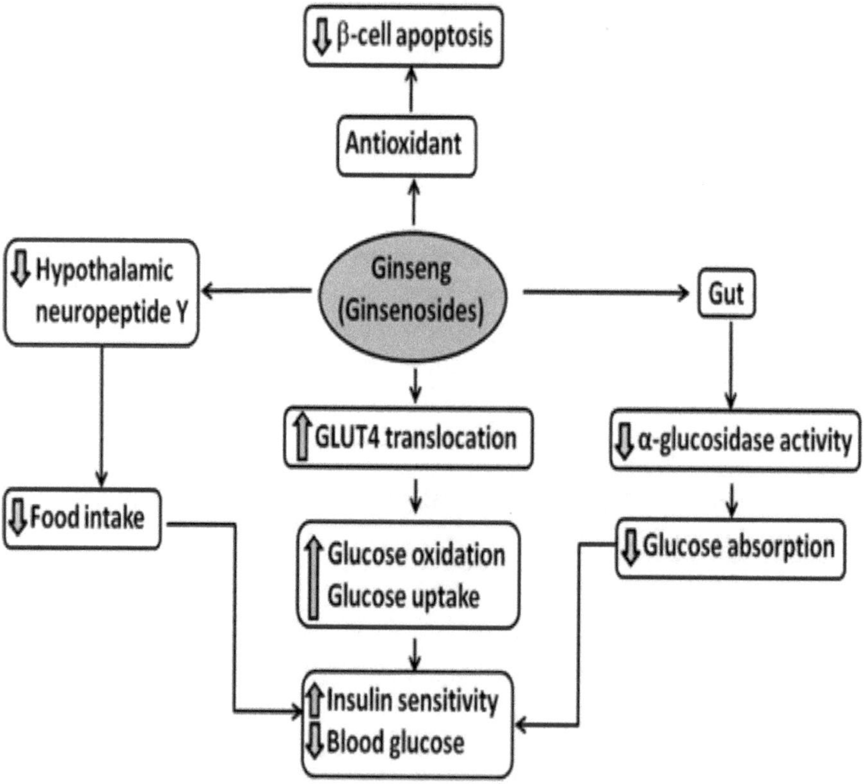

FIGURE 1.1 Mode of activity of ginseng to enhance glucose metabolism.
Source: Reprinted with permission from Ref.[68]. © 2011 Springer Nature.

1.2.3 FENUGREEK (TRIGONELLA FOENUM-GRAECUM)

Trigonella foenum-graecum is natural remedies source from dried seeds of a plant found in India, Africa, and China. It consists of active chemical constituent like 4-hydroxy isoleucine (an amino acid derived from fenugreek). It helps to enhance the production of insulin, reduce insulin resistance and therefore decrease blood glucose concentration. There is no more evidence that is available related to fenugreek for their anti-diabetic activity. *Trigonella foenum-graecum* when administered to STZ-induced diabetic rats leads to decrease blood glucose by 14.4 and 46.6%. It contains major alkaloids such as, lupinidine, lupanine, nicotinic acid, coumarine, and scopoletin [11–13].

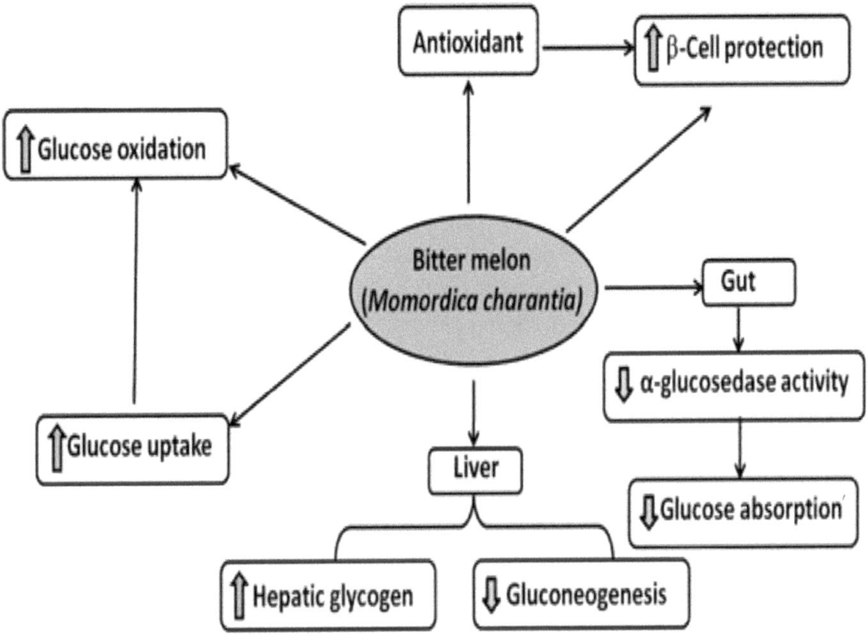

FIGURE 1.2 Diagrammatic pathway of bitter melon to reduce blood glucose.

Source: Reprinted with permission from Ref.[68]. © 2011 Springer Nature.

1.2.4 GYMNEMA

Gymnema, in India, is widely called "gurmar" which means 'sugar destroying.' The hypoglycemic activity of Gymnema can be tested along with insulin and results are encouraging. Research suggests that leaf extract of *G. sylvestre* containing gymnemic acid that suppresses the elevated blood sugar concentration by blocking glucose utilization in the intestine. The extract of the plant acts to work by increasing endogenous insulin production by increasing serum C-peptide levels. It is a chain of amino acid that is formed by cleaving of proinsulin molecules produces from the pancreas to form insulin. Moreover, Gymnema also decreases the level of cholesterol, triglyceride, and LDL-C, whereas it increases the level of HDL-C [12–14]. The gymnemic acid possesses hypoglycemic activity due to delay the glucose absorption in the blood. As gymnemic acid molecules fill the receptor location in the absorptive external layer of the intestine, therefore, absorption of sugar molecules inhibited, which result in low blood sugar level.

1.2.5 COPTIS CHINENSIS

Coptis chinensis is implied to manage diabetes in china. The active ingredient, berberine is an isoquinoline alkaloid present in *Coptis chinensis.* It is mainly found in the plant parts such as bark, stems, roots, and rhizomes. There is multiple mechanism involved for stimulation induced by berberine. Berberine increases GLUT 4 translocation in adipocyte and myotubes (Figure 1.3, Table 1.1). It also increases AMPK activity results decrease in glucose-stimulated insulin secretion (GSIS) and palmitate potential insulin secretion in MIN6 cell and islets [13].

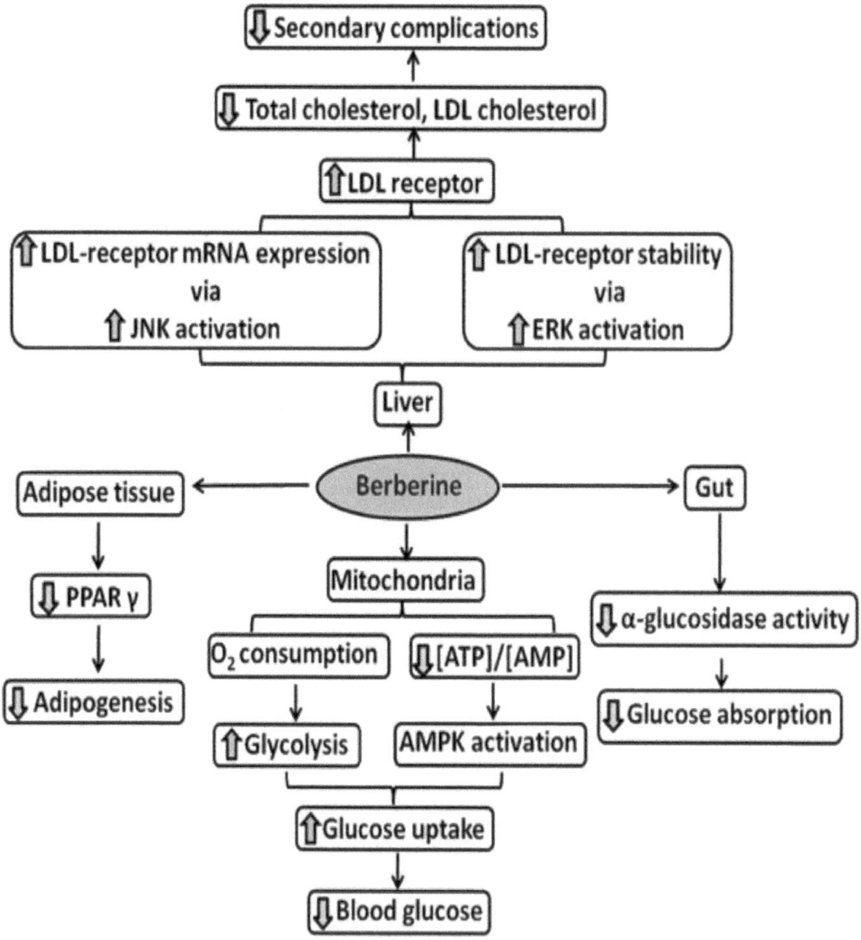

FIGURE 1.3 Mode of action of berberine in blood glucose regulation.

Source: Reprinted with permission from Ref.[68]. © 2011 Springer Nature.

TABLE 1.1 Some Common Plants and Their Mode of Action

Common Name	Mechanism of action
Babool	Release insulin from the pancreas
Sharifa	Increase plasma insulin
Mirch	Reduction of insulin binding on receptor site.
Dalchini	Increase plasma insulin
Eukalyptus	Increase insulin secretion from clonal pancreatic beta line
Shakarkand	Reduce insulin resistance
Chirayata	Increase insulin release at β-cells
Adrak	Facilitate insulin level
Sadabahar	Beta-cell regeneration and stimulation
Gurmar	Increase serum C-peptide level
Turmeric	Increase postprandial serum insulin
Soybean	Improve insulin secretion and increase β-cell mass
Indian Mulberry	Not yet established
Strawberry	Decrease gluconeogenesis
Onion	Increase glucose homeostasis

1.3 SOME COMMON PLANTS WITH MEDICINAL AND ANTI-DIABETIC PROPERTIES

1.3.1 STEVIA REBAUDIANA

It is commonly called as candy leaf, sugar leaf, and sweet leaf. *Stevia rebuadiana* Bertoni is a small branched bushy shrub that belongs to the aster or chrysanthemum family (Figure 1.4). It is mainly grown in mountain region of Paraguay. The principle bioactive constituent present in plants is stevioside, steviolbioside, and rebaudioside-A, B, C, D, E, F, and dulcoside. Other constituents are alkaloids, tannins, polyphenols, sugars (glucose, pentose, and fructose), flavonoids, and vitamins. *Stevia rebaudiana* Bertoni is considered as natural sweetener. It has anti-oxidant, anti-diabetic, and renal protective properties. It also possesses anti-hyperglycemic, insulinotropic, and glucagonostatic activity [20–24].

FIGURE 1.4 *Stevia rebaudiana* Bertoni leaves.

1.3.2 ALLIUM SATIVUM LINN

Allium sativum is a species of onion and is commonly called as Garlic, family Liliaceae (Figure 1.5). It is mainly incorporated in spicy flavoring agent since ancient times. It is mainly produced in Central Asia, Southern Europe, the USA, and India. It requires a well-drained climatic condition during growth. The bulb of garlic is white to pink in color having an aromatic or pungent taste. Oral ingestion of garlic extract and oil are very effective in lowering blood sugar concentration with effectiveness as compare closely to drug tolbutamide. Allicin (diallyl disulfide-oxide) and S-allyl cysteine sulfoxide (the precursor of garlic oil) are the main bioactive component which is important for anti-diabetic activity. It facilitates insulin production as well as increase liver glycogen synthesis. Other than the above bioactive component garlic also contains 29% carbohydrate, 56% protein (albumin), 0.1% fat and mucilage, and 0.1% volatile oil. A chief active constituent is allyl propyl disulfide, diallyl disulfide, alliin, and allicin. Raw garlic has a beneficial effect on reversing proteinuria as well as reducing blood sugar. The extraction of garlic in ethanol also shows significantly anti-diabetic activity. It is considered as an expectorant, stimulant, carminative, and antipyretic. Garlic also shows anti-bacterial, anti-fungal, and anti-hypertensive activity. It is also implied to treat eczema as well as in hyperlipidemia [25–29].

1.3.3 *CURCUMA LONGA LINN*

Turmeric is scientifically known as *Curcuma longa* Linn, family Zingiberaceae (Figure 1.6). It is also known as Indian saffron, Haldi, Curcuma, etc., in India. India contributed 90% of the total output of the world. Worldwide Turmeric is tremendously implemented as a spice, food preservatives, and coloring material. Scientific studies elaborated that the pharmacological properties of this spice (turmeric) and interest is mainly focused on curcumin as it is a chief bioactive component in turmeric. However, recent chromatographic studies show three curcuminoids such as curcumin, demethoxycurcumin (DMC), and bisdemethoxycurcumin (BDMC) which is always present together in turmeric sample. The chief component of *Curcuma longa* is called curcuminoids (curcumin and its analog), which is responsible for yellowish color in some species of curcuma. Turmeric also contains ascorbic acid, galantamine, and genistein. Other constituents such as 5% volatile oil, resin, and starch are present in turmeric. Basically, turmeric is taken as spices and coloring agents. Moreover, the curcuminoids have potential anti-oxidant properties which have a beneficial role in autoimmune disease such as type-1 diabetes mellitus, the supplementation of diabetic rats with turmeric or its active ingredients show attenuated hyperglycemia. Curcuminoids and its analog also have choleretic, hydrocholagogic, hypocholesterolemic, anti-inflammatory, anti-carcinogenic, and anti-oxidative properties [30–34].

FIGURE 1.5 *Allium sativum* Bulb.

1.3.4 *AEGLE MARMELOS CORR*

Bael is scientifically known as *Aegle marmelos* Corr. family Rutaceae. In Hindi indigenous to India, it is widely known as Bael. Bael is also said as a wood apple and is grown as a temple garden tree because leaves are considered to pray Lord Shiva. It is mostly used in Indian Ayurvedic medicine for the management of diabetes mellitus. *Aegle marmelos* is a native plant of India (Figure 1.7). The tree is medium-sized having armed deciduous mainly found in wild, easily found in a dry forest. The fruits are fleshy with soft, hard, and having aromatic rind. The ripped fruits are incorporated to treatment of stomachic and digestive complication. *Aegle marmelos* is also said to be traditional remedy for the both type of diabetic complication. The leaves of *Aegle marmelos* contain alkaloids, cardiac glycoside, terpenoids, flavonoids, tannins, saponin, and steroids. Moreover, fruits pulp contains lignin, terpenoids, flavonoids, phenolic compound, steroids, fat and oil, inulin, protein, and carbohydrate. Also, two novel compounds such as 2-iso-propenyl-4-methyl-1-oxa-cyclopenta[b]anthracene-5,10-dione and (+)-4-(2'-hydroxy-3'-enyloxy)-8*H*-[1,3]-dioxolo[4,5-H]chromen-8-one, are isolated from seeds of *Aegle marmelos* which give significantly *in-vitro* antifungal activity against pathogenic strain of fungi. Synthetic anti-hyperglycemic drugs such as Angeline, metformin fenofibrate are successfully separated from leaves of *A. marmelos*. The hypoglycemic effect of *Aegle marmelos* is probably attributed to its coumarine, which is responsible for the stimulation of insulin production from the pancreatic β-cell. The component of *Aegle marmelos* such as stem, fruits, leaves and roots have been considered as ethnomedicine for several medicinal properties: digestive, appetizer, astringent, anti-diarrheal, analgesics, demulcent, anti-pyretic, anti-dysenteric, hemostatic, anti-scorbutic, aphrodisiac and as an antidote for snakebite. Other than these activities of Bael, it also possesses anti-fungal, anti-microbial, anti-proliferative, hepatoprotective, cytoprotective, antifertility (it is reported with leaves of *Aegle marmelos* aqueous extract in albino male rats) and anti-oxidant activity [35–39].

FIGURE 1.6 *Curcuma longa.*

FIGURE 1.7 *Aegle marmelos.*

1.3.5 ANANAS COMOSUS

Medicinal plants possess various active moieties having therapeutic value and have been incorporated as remedies for humans since ancient times. There are numerous medicinal plants with potent anti-diabetic activity. In the series of medicinal plant, one such medicinal plants is *Ananas comosus*, widely called as Pineapple having Bromeliaceae family. Pineapple is a tropical fruit which grows in tropical and sub-tropical country, having worldwide production is 16–18 million tons. The plant is also famous for its Folk medicinal utilities, besides agricultural utilities being fruits for nutritional value. About 65% of fresh pineapple fruits is edible, which is composed with important nutrients, essential to human health. The scientific study explores that ethanolic extract of pineapple shows anti-diabetic as well as dyslipidemia activity.

The qualitative analysis of an ethanolic extract of *Ananas comosus* leaves shows that tannins, flavonoids, alkaloids, phytosterol, and glycosides are present (Figure 1.8). Pineapple also contains vitamins and sugar especially fructose, sucrose, and glucose. It is also observed that the pineapple pulp contains major volatile components such as terpenes, esters, ketones, and aldehyde. Some biomolecules having carbonyl, hydroxyl, and amine functional groups are also present having the potential for metal ion reduction.

Pineapple exists as the third most important tropical fruit having worldwide production is about 14 million tons. In Thailand, pineapple is considered to treat dysuria. In the Chinese medicine system, leaves of pineapple are incorporated into anti-dyspepsia and anti-diarrhea. The recent research study also proves that the pineapple and their part (fruits and leaf) show potential pharmacological activity like anti-diabetic, antibacterial, antioxidant, anti-dyslipidemia, and anti-tumor. Another series of studies show that the hypolipidemic effect is achieved due to the inhibition of HMG CoA. It is considered that the anti-diabetic and anti-dyslipidemia effect with their mechanism partially revealed and the profuse sources of pineapple leave [40–43].

FIGURE 1.8 Pineapple.

1.3.6 *OCIMUM SANCTUM LINN*

Tulsi is a wonderful herbal drug in the Indian system of traditional medicine which is also called in English "Holy Basil." As per biological nomenclature, Tulsi is called *Ocimum sanctum* Linn belong to family Lamiaceae (Figure 1.9). In India, Tulsi is also considered as a sacred plant in Hindu culture. It is a small herb (about 18 inches long) found in tropical wild cultivated annually. Basically, this plant is cultivated or grows in wild but is commonly grow in the home and temple garden in India. Apart from the religious significance, it also has a long history of medicinal use and is mentioned in *Charak samhita*, the ancient textbook of *Ayurveda*. Modern studies show that the leaves extract of Tulsi significantly possesses hypoglycemic properties in single-blind patients. *Ocimum sanctum* is a wonderful herbal drug having various pharmacological activity. Scientific study established the evidence of triterpenoids with hydroxyl or carboxylic or ketonic group which are significantly responsible for anti-diabetic activity. Triterpenoids reduce the serum glucose, tryglycerides, LDL, cholesterol, and total cholesterol. Tulsi also contains bioactive moiety other than triglycerides such as ursolic acid, eugenol, rosmarinic acid, caryophyllene, and oleanic acid. The seeds of tulsi contain fixed oil such as linoleic acid and linolenic acid.

The seed oil of Holy Basil is earlier reported for anti-proliferative activity against HeLa cells in culture. Also, the seed of *Ocimum sanctum* is reported for hypoglycemic, hypouricemic, anti-inflammatory, and analgesic activity. Tulsi is also called as "Queen of plants" because of its bio-pharmacological activity like anti-malarial, anti-diabetic, anti-inflammatory, anti-bacterial, anti-fungal, expectorants, insecticidal, anti-emetic, anti-stress, anti-lipid peroxidative, anti-cholesterol, anti-lipidemic, anti-ulcer, and immunomodulatory properties. Stem extract and a leaf of Tulsi significantly show anti-convulsant and anti-depressant activity simultaneously [44–48].

FIGURE 1.9 *Ocimum sanctum.*

1.3.7 PIPER BETLE LINN

The leaves of Piper betle Linn have long been used in the Indian traditional system of medicine and is widely said to be "*Paan*" in Assamese, Urdu, Hindi, Odia, and Bengali. In Sanskrit, it is far known as Tambula and Nagavalli. It belongs to family Piperaceae. In ancient India, it is considered that paan is discovered with the help of Dhanvantary by the scholars of Ayurveda, thousands of years ago. Evidence of Paan is also found in Shrimad Bhagavatam as Lord Krishna used to chew it, but the tradition of chewing Paan was famous by Queen Noorjehan, Empress of the Mughal Dynasty. Betal leaves are also considered in Hindu culture and are still extensively used while religious functions in Hindu culture. It is mainly grown in the moist climatic regions like East India (Bihar, Bengal, and Orissa), South India, and Karnataka. *Piper* species are most popularly used in traditional medicine systems such as Chinese medicine, Ayurveda, and Folklore medicine of West Indies and Latin America.

The *P. betle* leaves reported to yield an alkaloid arakene. Chief components of the leaves is volatile oil, 0.9 to 1.8% chavicol, betelphenol, eugenol 0.32%, allyl pyrocatechin 0.71%, terpene, cineol 0.04%, caryophyllene 3.71%, and menthone. *Piper betle* also contain chavibetol 53.1%, chavibetol acetate 15.5%, chavibetol methyl ether 0.48%, campene 0.48%, f-pinene 0.21%, u-limonene 0.14%, and saprobe 0.11%. Other than chemical constituent, elemental composition are present *P. betle* such as water 85–90%, protein 3%, fat 1%, mineral 3%, energy 44 kcal/100g, essential oil 0.2%, tannins 0.1–0.3%, vitamin A 2%, vitamin C 0.01%, Nitrogen 2–7%, and phosphorus 0.05–0.6%, respectively.

FIGURE 1.10 *Piper betle* plant.

The effect of *Piper betle* on fasting blood glucose levels were reported by the scientific study (Figure 1.10). It was found that the *P. betle* are very effective in reducing significantly blood glucose concentration upto 4 hours except lowest dose, which impaired blood glucose concentration only upto 2 hours. Also, leaves of *Piper betle* shows anti-oxidant, gastroprotective, anti-bacterial, anti-fungal, anti-diabetic, anti-inflammatory, radioprotective, hepatoprotective, anti-fertility in male rats, anti-motility and wound healing. Paan is also used to chewing as a mouth freshener, digestive, stimulant, and carminative. *P. betle* is used for curing wind-cold cough, bronchial asthma, rheumatism, and pregnancy edema in the Chinese folk medicine system [49–53].

1.3.8 MURRAYA KOENIGII

Curry leaf or Curry patta is botanically known as *Murraya koenigii* belongs to family Rutaceae (Citrus family). It is widely incorporated as spice, condiment and flavoring agent in Indian food and dishes. It is also reported to use as an anti-diabetic agent by Ayurveda. The leaf extract significantly reduces the concentration of blood glucose in experimental diabetic animals and carbazoles from *Murraya koenigii* leaf extract possess strong anti-oxidant activity (Figure 1.11). It is an unarmed semi-deciduous shrub or small tree having a woody stem. The plant is cultivated and grown throughout India. It is found wild of Himalaya, Uttarakhand, Sikkim, Bengal, Assam, and the Western Ghats. It is also cultivated in the moist region of Asia like Bhutan, Nepal, Pakistan, Sri Lanka, and Thailand.

An active chemical constituent called Mahanimbine from *M. koenigii* is isolated by column chromatography of petroleum ether extract are reported to show anti-diabetic activity. Other bioactive constituents are also present in Curry patta such as koenimbine, murrayacine, girinimbine, koenidine, koenine, koenigine, mukonicine, mahanine, carotene, mahanimbinine, mahanimbicine, mahanimboline, isomahanine, mukoeic acid, lutein, murrayanine, mukonine, isomurrayazoline, murrayazolinine, isomurrayazolinine, mukonidine, mukoline, mukolidine, girinimbilol, koenoline, and glycozoline. *In-vivo* pharmacological studies of *Murraya koenigii* show vasodilating, anti-diabetic, hypocholesterolemic, anti-ulcer, anti-diarrheal, phagocytic, analgesic, anti-nociceptive, antilipid peroxidative, radio-protective, chemo-protective, anti-amnesic, anti-helminthic, memory enhancing, and wound healing activity. *In-vitro* study of *Murraya koenigii* shows anti-microbial, anti-oxidative, skin pigmenting, cytotoxic, and anti-tumor activity [54–58].

FIGURE 1.11 *Murraya koenigii* along with fruit and flowers.

1.3.9 AZADIRACHTA INDICA

Azadirachta indica is commonly known as Neem and it belongs to family Mahogany/Meliaceae. It is widely used in treatment of multiple acute and chronic diseases in various parts of the world including Asia and Africa. Neem is the very first medicinal plant mentioned in the Siddha medicine system, which is the oldest medical system known to mankind, having origin back to 10,000 BC to 4000 BC. It is also considered as the most respected tress in among Indians due to its wonderful uses. "Arista" is a Sanskrit word given to the Neem that means "Perfect" complete and imperishable. Arishtha is Sanskrit's name for Neem tree meaning 'reliever of sickness' and hence, it is considered as "Sarbarogaribarini." The Neem trees are mainly found in India, East and sub-Sahelian Africa, South America and Southeast Asia. Neem is called as 'storehouse' for most of phytochemicals. Till date, over 300 phytochemicals have been isolated from neem. These chemicals are found as structurally very complex and chemically very diverse. The leaves of neem contain 0.13% essential oil which gives smell to leaves. Mainly two major classes of phytochemicals are present in the neem such as isoprenoids and non-isoprenoids which are isolated from different parts of tree. The most common isoprenoids are diterpenoids, triterpenoids, vilasinine, limonoids, and C-secomeliacine. The non-isoprenoids include protein, polysaccharides, a sulfur compound, polyphenolics, dihydrochalcone, coumarins, tannins, and aliphatic compounds. Neem also contains nimbidin, sodium nimbidate, azadirachtin, nimbin, nimbolide, gedunin, gallic acid, mahmoodin, and margolone. Several herbal components and drugs have been used to control diabetes mellitus, in the series of herbal component neem is playing a beneficial role to manage diabetes mellitus. In 1986, the scientific study demonstrated the anti-hyperglycemic and hypoglycemic activity of neem seed oil in alloxan-induced diabetes female rats. Neem also shows important biological activity such as an anti-inflammatory, anti-arthritic, antipyretic, hypoglycemic, anti-gastric ulcer, spermicidal, anti-microbial, diuretic, anti-malarial, immune-modulator, anti-tumor, CNS depressant, antifertility, and anti-tuberculosis [59–64].

1.3.10 ALOE BARBADENSIS MILLER

The biological source of *Aloe vera* is *Aloe barbadensis* belonging to family Liliaceae. Liliaceae is indigenous to dry sub-tropical and tropical climate but it originated in South Africa and southern USA. It is folk medicine for over 2000years and have great importance in traditional medicine of many countries like China, India, West Indies, and Japan. Aloe vera develops water storing tissue in the leaves to survive in the dry and low rainfall area because of its succulent xerophyte properties. The leaves of Aloe vera basically divide into parts such as outer green rind and inner colorless parenchyma. The outer rind contains vascular bundle and inner part contains aloe gel. Aloe vera show marvelous medicinal activity. It reduces blood sugar levels in diabetic patients also it improves the responsiveness of body tissue towards insulin. Hence, Aloe vera more suitable for wide range treatment to all type of diabetic complication.

The bio-active component of *Aloe vera* is barbaloin including anthraquinones, chromones, polysaccharides, and enzymes. The other active principles present in outer layer of leaf are hydroxyanthrone, aloe emodin-anthrone 10-C glucoside. The mid-layer of parenchyma contains protein, lipid, amino acid, vitamins, enzymes, inorganic compound, small organic compound, and carbohydrate. The inner layer of leaf contains vitamin, enzymes, minerals, sugars, lignin, saponins, salicylic acid, and amino acid. The potential therapeutic use of Aloe barbadensis is wound healing, Moisturizing and anti-aging agent, anti-inflammatory, anti-diabetic, anti-microbial, anti-septic, anti-viral, anti-oxidant, anti-hyperlipidemic, anti-ulcer, anti-tumor, laxative effect. The medical uses of *Aloe vera* include anthelmintic, aperient, carminative, depurative, diuretic, stomachic, burns, dyspepsia, amenorrhea, hyperadenosis, hepatopathy, constipation, and abdominal tumor [65–67].

1.4 CONCLUSION

The natural medicinal products possess many pharmacological activities like anti-diabetic, anti-ulcer, anti-tumor, anti-pyretic, anti-lipidemic, wound healing, etc. The natural products are widely used from therapeutic purpose from ancient time. Many of the natural products have a well-known mechanism for their activity but still, most of the natural products have not established mechanisms as well as an active therapeutic component for their activity, such type of natural products are mainly incorporated into work to establish their bioactive component by using a sophisticated instrument such as ultraviolet spectroscopy, mass spectroscopy, NMR, etc. Natural products also contain some wonderful drugs such as Tulsi, Neem, and Aloe vera, which possess great potential to therapeutic uses from the ancient times of mankind but still the mechanism some pharmacological activity of these plants is not known. In the future, a controlled study is required to prove the effectiveness of natural products under various conditions.

KEYWORDS

- **chemical constituents**
- **diabetes**
- **herbal medicine**
- **natural product**
- **treatment**

REFERENCES

1. Mishra, V., Mishra, S. K., et al., (2014). Antiulcer potential of leaves of *Achyranthes aspera Linn* in rats. *Natural Product an Indian Journal, 10*(4), 123–129.
2. Ahmed, L., Semotiuk, A. J., et al., (2018). Anti-hypertensive plants of rural Pakistan: Current use and future potential. *Journal of Complementary Medicine Research, 7*(2), 138–153.
3. Nikhade, N., Ansari, M., et al., (2019). A review of natural antioxidant in medicinal plants. *International Journal of Pharmaceutics and Drug Analysis, 7*(2), 11–15.
4. Osadebe, P. O., Odoh, E. U., et al., (2014). Natural product as potential source of anti-diabetic drugs. *British Journal of Pharmaceutical Research, 4*(17), 2075–2095.

5. Mahapatra, D. K., Asati, V., et al., (2015). Chalcones and their therapeutic target for the management of diabetes: Structural and pharmacological perspectives. *European Journal of Medicinal Chemistry, 92*, 839–865.
6. Upadhyay, R. K., (2016). Anti-diabetic potential of plant natural products: A review. *International Journal of Green Pharmacy, 10*(3), 1–2.
7. Chauhan, A., Sharma, P. K., et al., (2010). Plant having potential anti-diabetic activity: A review. *Der Pharmacia Lettre, 2*(3), 369–387.
8. Bhat, M., Zinjarde, S. S., et al., (2008). *Anti-diabetic Indian Plants: A Good Source of Potent Amylase Inhibitors* (pp. 1–6). Hindawi publishing corporation, Evidence based complementary and alternative medicine.
9. Doble, M., et al., (2011). Mechanism of action of natural products used in the treatment of diabetes mellitus. *Chin. J. Integr. Med., 17*(8), 563–574.
10. Bhushan, M. S., Rao, C. V., et al., (2010). *An Analytical Review of Plants for Anti-Diabetic Activity with Their Phyto-Constituent and Mechanism of Action. Int J. Pharm. Sci. Res.1*(1), 29–46.
11. Pereg, G R. M., Zavala, S M. A., et al., (1998). Anti-diabetic effect of compounds isolated from plant. *Phytomedicine, 5*(1), 55–75.
12. Mentreddy, S. R., (2007). Medicinal plant species with potential antidiabetic properties. *Journal of the Science of Food and Agriculture, 87*, 743–750.
13. Doble, M., et al., (2011). Mechanism of action of natural products used in the treatment of diabetes mellitus. *Chin. J. Integr. Med., 17*(8), 563–574.
14. Rao, M. U., Sreenivasulu, M., et al., (2010). Herbal medicine for diabetes mellitus: A review. *International Journal of Pharm. Tech. Research, 2*(3), 1883–1892.
15. Patel, D. K., Prasad, S. K., et al., (2012). An overview on anti-diabetic plants having insulin mimetic properties. *Asian Pacific Journal of Tropical Biomedicine*, 320–330.
16. Doble, M., et al., (2011). Mechanism of action of natural products used in the treatment of diabetes mellitus. *Chin. J. Integr. Med., 17*(8), 563–574.
17. Rios, J. S., Francini, F., et al., (2015). Natural product for the treatment of type 2 diabetes mellitus. *Planta. Med., 81*, 975–994.
18. Olajide, O. A., Olubusayo, A. W. E., et al., (1999). Evaluation of anti-diabetic property of *Morinda lucida* leaves in streptozotocin diabetic rat. *J. Pharm Pharmacol., 51*, 1321–1324.
19. Vinayagam, R., et al., (2015). Anti-diabetic properties of dietary flavonoids: A cellular mechanism review. *Nutrition and Metabolism, 12*(60), 1–20.
20. Shivanna, N., Naika, M., et al., (2013). Antioxidant, anti-diabetic and renal protective properties of *Stevia rebaudiana. Journal of Diabetes and its Complication., 27*, 103–113.
21. Abou-Arab, E., Abou-Arab, A., et al., (2010). Physicochemical assessment of natural sweeteners steviosides produced from *Stevia rebaudiana* bertoni plant. *African Journal of Food Science, 4*(5), 269–281.
22. Lemus-Mondaca, R., Zura-Bravo, L., et al., (2012). *Stevia rebaudiana* Bertoni, source of a high potency natural sweetener: A comprehensive review on the biochemical, nutritional and functional aspects. *Food Chemistry, 132*, 1121–1132.
23. Mamta, Rahi, P., et al., (2010). Stimulatory effect of phosphate solubilizing bacteria on plant growth, stevioside and rebaudioside-A content of *Stevia rebaudiana* Bertoni. *Applied Soil Ecology, 46*, 222–229.
24. Ramesh, K., Singh, V., Megeji, N. W., (2006). Cultivation of stevia [Stevia rebaudiana (Bert.) Bertoni]: A comprehensive review. *Advances in Agronomy. 89*, 137-77.

25. Malvi, R., Jain, S., et al., (2011). A review on anti-diabetic medicinal plants and marketed herbal formulations. *International Journal of Pharmaceutical and Biological Archives, 2*(5), 1344–1355.
26. Kokate, C. K., Gokhale, S. B., et al., (2005). *A Textbook of Pharmacognosy* (32nd edn., pp. 347–348). Nirali prakashan.
27. Srivastava, P., Kumar, N., et al., (2010). Plant having potential anti-diabetic activity: A review. *Scholar Research Library,* 2(3), 369–387.
28. Frode, T. S., et al., (2008). Animal model to test drugs with potential anti-diabetic activity. *Journal of Ethnopharmacology, 115*, 173–183.
29. Eidi, A., Eidi, M., et al., (2006). Anti-diabetic effect of garlic (*Allium sativum* L.) in normal and streptozotocin-induced diabetic rats. *Phytomedicine, 13*, 624–629.
30. Kokate, C. K., Gokhale, S. B., et al., (2019). *A Textbook of Pharmacognosy* (65th edn., pp. 427–433). Nirali prakashan.
31. Kalaycioglu, Z., et al., (2017). Comparison of anti-oxidant, anti-cholinesterase, and anti-diabetic activities of three curcuminoids isolated from *Curcuma longa* L. *Natural Product Research, 31*(24), 2914–2917.
32. Kumar, S., Kumar, M., et al., (2011). An aqueous extract of *Curcuma longa* (turmeric) rhizomes stimulates insulin action on tissue involved in glucose homeostasis *In-Vitro. Phytotherapy Research, 25*, 396–401.
33. Pari, L., Saravanan, G., et al., (2002). Anti-diabetic effect of cogent db, a herbal drug in alloxan induced diabetes mellitus. *Comparative Biochemistry and Physiology, 131*, 19–25.
34. Lekshmi, P. C., Nisha, V. M., et al., (2013). In-vitro anti-diabetic and inhibitory potential of turmeric (*Curcuma longa* L.) rhizome against cellular and LDL oxidation and angiotensin converting enzyme. *Journal of Food Science Technology.*
35. Kesari, A. N., Gupta, R. K., et al., (2006). Hypoglycemic and anti-hyperglycemic activity of *Aegle marmelos* seed extract in normal and diabetic rats. *Journal of Ethno-Pharmacology, 107*, 374–379.
36. Sekar, D. K., Kumar, G., et al., (2011). A review on pharmacological and phytochemical properties of *Aegle marmelos* (L.) Corr. Serr. (Rutaceae). *Asian Journal of Plant Science and Research, 1*(2), 08–17.
37. Mishra, B. B., Singh, D. D., et al., (2010). Antifungal constituent isolated from the seed of *Aegle marmelos. Phytochemistry, 71*, 230–234.
38. Narendra, T., Shweta, S., et al., (2007). Anti-hyperglycemic and anti-dyslipidemic agent from *Aegle marmelos. Bio-Organic and Medicinal Chemistry Letters, 17*, 1808–1811.
39. Singab, A. N., et al., (2014). Medicinal plant with potential anti-diabetic activity and their assessment. *Medicinal and Aromatic Plants, 3*(1), 1–12.
40. Kalpana, M. B., Subramanian, S., et al., (2014). Studies on the anti-diabetic activity of *Ananas comosus* leaves in STZ induced diabetic rats. *Der Pharmacia Letter, 6*(1), 190–198.
41. Riya, M. P., Antu, K. A., et al., (2014). An in-vitro study reveals nutraceutical properties of *Ananas comosus* (L.) Merr. Var. *Mauritius* fruits residue beneficial to diabetes. *Journal of Science Food Agriculture, 94*, 943–950.
42. Emeka, E. E., Ojiefoh, O. C., et al., (2013). Evaluation of antibacterial activity of silver nanoparticles green-synthesized using pineapple leaf. *Micron,* 01–05.
43. Hu, J., Lin, H., et al., (2011). Developmental toxicity of orally administered pineapple leaf extract in rats. *Food and Chemical Toxicity, 49*, 1455–1463.

44. Umesha, S., Marahel, S., et al., (2013). Antioxidant and anti-diabetic activities of medicinal plant: A short review. *International Journal of Research in Phytochemistry and Pharmacology, 3*(1), 40–53.
45. Vats, V., Yadav, S. P., et al., (2004). Ethanolic extract of *Ocimum* sanctum leaves partially attenuates streptozotocin-induced alteration in glycogen content and carbohydrate metabolism in rats. *Journal of Ethno-Pharmacology, 90*, 155–160.
46. Patil, R., Ashirwar, B., et al., (2011). Isolation and characterization of anti-diabetic component (bioactivity guided fractionation) from *Ocimum sanctum* L. (Lamiaceae) aerial part. *Asian Pacific Journal of Tropical Medicine*, 278–282.
47. Sethi, J., Sood, S., et al., (2004). Evaluation of hypoglycemic and antioxidant effect of *Ocimum sanctum. Indian Journal of Clinical Biochemistry, 19*(2), 152–155.
48. Prakash, J., Gupta, S. K., et al., (2000). Chemopreventive activity of *Ocimum sanctum* seed oil. *Journal of Ethno-Pharmacology, 72*, 29–34.
49. Shah, S. K., Garg, G., et al., (2016). *Piper betle*: Phytochemical, pharmacological and nutritional value in health management. *International Journal of Pharmaceutical Science Review and Research, 38*(2), 181–189.
50. Rekha, V. P. B., Kollipara, M., et al., (2014). A review on *Piper betle L.*: Nature's promising medicinal reservoir. *American Journal of Ethnomedicine, 1*(5), 276–289.
51. Arambewela, L. S. R., Arawwawala, L. D. A. M., et al., (2005). Anti-diabetic activity of aqueous and ethanolic extracts of *Piper betle* leaves in rats. *Journal of Ethno-Pharmacology, 102*, 239–245.
52. Wirotesangthong, M., et al., (2008). Inhibitory effect of *Piper betle* on production of allergic mediators by bone marrow-derived mast cell and lung epithelial cells. *International Immune-Pharmacology, 8*, 453–457.
53. Punuri, J. B., Sharma, P., et al., (2012). *Piper betle* mediated green synthesis of biocompatible gold nanoparticles. *International Nano Letters, 2*(18), 01–09.
54. Kesari, A. N., Gupta, R. K., et al., (2005). Hypoglycemic effect of *Murraya koenigii* on normal and alloxan-diabetic rabbits. *Journal of Ethno-Pharmacology, 97*, 247–251.
55. Yadav, S., Vats, V., et al., (2002). Hypoglycemic and anti-hyperglycemic activity of Murraya koenigii leaves in diabetic rats. *Journal of Ethno-Pharmacology, 82*, 111–116.
56. Arulselvan, P., et al., (2007). Beneficial effect of *Murraya koenigii* leaves on anti-oxidant defense system and ultra-structural changes of pancreatic β-cells in experimental diabetes in rats. *Chemico-Biological Interaction, 165*, 155–164.
57. Saini, S. C., et al., (2015). A review on curry leaves (*Murraya koenigii*): Versatile multi-potential medicinal plant. *American Journal of Phyto-Medicine and Clinical Therapeutics, 3*(4), 363–368.
58. Handral, H. K., et al., (2012). A review on *Murraya koenigii*: Multi-potential medicinal plant. *Asian Journal of Pharmaceutical and Clinical Research, 5*(4), 05–14.
59. Gupta, S. C., Prasad, S., et al., (2017). Neem (*Azadirachta indica*): An Indian traditional panacea with modern molecular basis. *Phytomedicine, 34*, 14–20.
60. Kumar, V. S., Navaratnam, V., et al., (2013). Neem (*Azadirachta indica*): Prehistory to contemporary medicinal uses to humankind. *Asian Pacific Journal of Tropical Bio-Medicine, 3*(7), 505–514.
61. Sharma, P., Tomar, L., et al., (2011). Review on neem (*Azadirachta indica*): Thousand problem one solution. *International Research Journal of Pharmacy, 2*(12), 97–102.

62. Dallaqua, B., Saito, F. H., et al., (2012). Treatment with *Azadirachta indica* in diabetic pregnant rats: Negative effects on maternal outcome. *Journal of Ethnopharmacology, 143*, 805–811.
63. Patil, P., Patil, S., et al., (2013). Anti-diabetic activity of alcoholic extract of neem (*Azadirachta indica*) root bark. *National Journal of Physiology, Pharmacy and Pharmacology, 3*(2), 142–146.
64. Boeke, S. J., Boersma, M. G., et al., (2004). Safety evaluation of neem (*Azadirachta indica*) derived pesticides. *Journal of Ethno-Pharmacology, 94*, 25–41.
65. Maharajan, H. R., et al., (2015). Evaluation of biological properties and clinical effectiveness of *Aloe vera*: A systematic review. *Journal of Traditional and Complementary Medicine, 5*, 21–26.
66. Choudhari, M., Kochhar, A., et al., (2011). Hypoglycemic and hypolipidemic effect of *Aloe vera* L. in non-insulin dependent diabetes. *Journal of Food and Science Technology*, 1–7.
67. Sahu, P. K., Giri, D. D., et al., (2013). Therapeutic and medicinal uses of *Aloe vera*: A review. *Pharmacology and Pharmacy, 4*, 599–610.
68. Prabhakar P.K., Doble M. Mechanism of action of natural products used in the treatment of diabetes mellitus. Chin J Integr Med. 2011 Aug;17(8):563-74. doi: 10.1007/s11655-011-0810-3. Epub 2011 Aug 9. PMID: 21826590.

CHAPTER 2

Exploring the Hidden Ethnobotanical Treasures for Treating or Managing Chronic Alcoholism

SAMEER A. HEDAOO[1] and DEBARSHI KAR MAHAPATRA[2]

[1]*Department of Pharmaceutical Chemistry, Gurunanak College of Pharmacy and Technical Institute, Nagpur – 440026, Maharashtra, India*

[2]*Assistant Professor, Department of Pharmaceutical Chemistry, Dadasaheb Balpande College of Pharmacy, Nagpur – 440037, Maharashtra, India, E-mail: mahapatradebarshi@gmail.com*

ABSTRACT

Alcohol is one of the most common beverages consumed in today's society. Its vast popularity at various levels makes it one of the highest causes of driving fatalities, especially in the United States. Reports and statistics state that over 65 million people are struggling with alcoholism disorder. Alcoholism is a medical, social and economic problems where treatment methods mostly include difficult and long-lasting psychotherapy and in some cases, quite controversial pharmacological approaches. No medicine can 'cure' alcohol use disorder, but some can help you to recover. They can make drinking less enjoyable so you don't want to do it as such; e.g., disulfiram, acamprosate, naltrexone, etc. But there are various side effects of synthetic drugs such as excessive use of acamprosate may results in abdominal pain, confusion, and increase urination. Naltrexone results stomach cramping, anxiety, etc. Excessive use of disulfiram shows polyneuritis, hepatitis, acneiform eruption, etc. The developments of alcoholism remedies have medical, social, and economical significance. In view of the adverse behavioral effects of synthetic drugs, the development of low toxicity and high efficiency medicines derived from natural products exhibits expansive market prospects. A number of medicinal plants and pure natural compounds are reported to have preventive and therapeutic effect on alcoholism and alcohol dependency, but their constituents, efficacy, and mechanism of action are mostly unknown so far. The reduction of alcohol absorption from the gastrointestinal system appears to be a common feature among most of the plants. The aim of this book chapter is to provide an overview of effective natural products for the prevention and treatment of chronic alcohol-use disorder.

2.1 INTRODUCTION

2.1.1 ALCOHOL ADDICTION AND ALCOHOLISM

Alcoholism is a disease for which counseling and rehabilitation for alcohol addiction or alcohol abuse is recommended. If you ignore the problem, it will become destructive physically, emotionally, mentally, and socially, tearing apart your relationships as it controls your mind and body [1]. Alcohol addiction is simply defined as a compulsive need for an intoxicating liquid that is obtained from fermented grain or fruits. These liquids include beer, wine, and other hard liquors. Alcohol is an addictive drug. The more alcohol we drink (and abuse) the more our body comes to depend on it -thus, the nature of addiction [2].

2.1.2 NEUROLOGICAL EFFECT OF ALCOHOL

GABAAR and ADH are important pharmacological concern in alcoholism. Alcohol consumption caused the changes in level of several GABAAR accompanied by behavioral disorder, e.g., loss of righting reflex. The primary pathway of alcohol metabolism cause oxidation to acetaldehyde catalyzed by ADH and followed by faster oxidation to acetate catalyzed by ALDH; i.e., ADH is one of the most important enzymes for decreasing alcohol concentration in the body [3].

2.1.2.1 TYPES OF ALCOHOLISM

Alcohol dependence was earlier termed as "alcoholism." The term alcoholism was however dropped later as it was considered to be vague and had a negative connotation. Magnus Huss was the first person to systematically classify the damage that was attributable to alcohol ingestion. The concept and classification of alcohol dependence has undergone a series of changes over the decades, which can be summarized under [4]. The Jellinek's classification of alcoholism and Cloninger's classification of alcoholism are depicted in Tables 2.1 and 2.2, respectively.

TABLE 2.1 Jellinek's Classification of Alcoholism

Class	Features
Alpha alcoholism	Excessive and inappropriate drinking to relieve physical and/or emotional pain. There was no loss of control and the ability to abstain was present.
Beta alcoholism	Excessive and inappropriate drinking, with physical complications like cirrhosis due to cultural drinking patterns and poor nutrition. No dependence.
Gamma (malignant) alcoholism	A progressive course marked by physical dependence with tolerance and withdrawal symptoms as well as psychological dependence with an inability to control drinking.
Delta alcoholism	An inability to abstain with presence of tolerance and withdrawal symptoms. The amount of alcohol consumed can be controlled and there is minimal social disruption.
Epsilon alcoholism	Dipsomania or compulsive drinking and spree drinking.

TABLE 2.2 Cloninger's Classification of Alcoholism

Class	Features
Type I alcoholism	Occurred in both sexes, with age of onset less than 25 years, possible family history of alcoholism and strong influence of both genetic and environmental factors. It was marked with psychological dependence and presence of guilt. The pre-morbid personality traits were harm avoidance and reward dependence.
Type II alcoholism	Limited to males, with age of onset older than 25 years and possibility of parental alcoholism and antisocial behavior. Genetic and environmental influences were limited. Additional features included no loss of control, drinking followed by aggressive behavior and spontaneous alcohol seeking. pre-morbid personality traits of novelty seeking were usually present.

2.2 TREATMENT PRINCIPLES AND OVERVIEW

Pharmacotherapy should be used in patients with alcohol use disorder such as disulfiram, naltrexone, and acamprosate. These drugs are most commonly used for alcoholism but they show various side effects like polyneuritis, hepatitis, acneiform eruption, skin rash, drowsiness, tiredness, headache, chest pain, hallucination, etc. Because of these side effects, the world is moving towards the natural therapy. The therapeutic effects of these natural remedies may be less pronounced several-times than synthetic drugs, but the probability of adverse symptoms to minimum [5, 6].

2.3 HERBAL DRUGS FOR TREATING CHRONIC ALCOHOLISM

2.3.1 ALOE VERA

- **Synonym:** Gheekumari (Hindi), Khorpad (Marathi).
- **Biological source:** Aloe vera commonly referred as *Aloe barbadensis* belonging to the family *Liliaceae.*
- **Chemical constituents:** The aloe pulp was shown to contain proteins, lipid, amino acids, vitamins, enzymes, inorganic compound, and small organic compound with addition to the different carbohydrate. Aloe vera gel polysaccharides consist of linear chains of glucose and mannose molecules.
- **Pharmacological uses:** Various extract of this aloe used to treat arthritis, skin cancer, burns, eczema, psoriasis, digestive problems, inflammation, high blood pressure, and diabetes.
- **Use in alcoholism:** Liver is the first organ which affected with the alcohol abuse. So, the recovery and healing must be started with healing the liver. Aloe vera effectively cleanses and detoxifies the body. Intake of a glass of Aloe vera juice everyday helps in fighting against alcoholism [7].

2.3.2 AMLA

- **Synonym:** Indian gooseberry (English), Amlika, Amba, Aamlah (Hindi).
- **Biological source:** It consists of ripe fruits of *Emblica officinalis* belonging to family *Euphorbiaceae.*
- **Chemical constituents:** It contains alkaloids, amino acids, mucic acid, tannins, flavones glycosides, phenolic glycoside, flavonols, phenolic acid, sesquiterpenoid, norsesquiterpenoid, and carbohydrates.
- **Pharmacological uses:** It prevents hyperlipidemia, osteoporosis, and several other ailments. *E. officinalis* supplementation offsets

N-nitrosodiethylamine (NDEA)-induced liver injury, *via* its anti-oxidant and anti-inflammation effects.

- **Use in alcoholism:** It prevents inflammation and oxidative stress occurred due to alcohol contribute to liver injury. *E. officinalis* protects against alcohol (hepatoxicity)-induced liver injury [8].

2.3.3 ANGELICA

- **Synonym:** Wild celery (English), Garden angelica, Norwegian angelica.
- **Biological source:** It consists of leaves of *Angelica archangelica* belonging to family *Apiaceae*.
- **Chemical constituents:** It consists of α-pinene, β-pinene, camphene, β-phellandrene, limonene, borneol, carvone, and caryophyllene.
- **Pharmacological uses:** It is used in heartburn, intestinal gas, loss of appetite, arthritis, etc.
- **Use in alcoholism:** Angelica plant has detoxifying properties, which reduces the cravings for alcohol. It makes the taste of alcohol disgusting and creates a disliking for it. It also has an anti-histamine property which is used for driven and compulsive drinkers [9].

2.3.4 APPLE

- **Synonym:** Crabapple (English), Seb (Hindi).
- **Biological source:** It is an edible fruit produced by tree *Malus domestica* belonging to family *Rosaceae*.
- **Chemical constituents:** It consist of vitamin A, B_1, B_2 and B_6, niacin, pantothenic acid, folic acid, copper, iron, magnesium, and zinc.
- **Pharmacological uses:** Trees and fruits are prone to a number of fungal, bacterial, and pest problems.
- **Use in alcoholism:** Addicts should take it at regular intervals. This helps to decrease alcohol craving and to remove the toxins from the body by enhancing the urine output [10].

2.3.5 ARJUNA

- **Synonym:** Arjun tree (English), Thella maddi (Telugu), Hole matthi (Kannada).

- **Biological source:** It is obtained from the tree *Terminalia arjuna* belonging to family *Combretaceae.*
- **Chemical constituents:** It contains arjunic acid, arjunolic acid, arjungenin, gallic acid, and ellagic acid.
- **Pharmacological uses:** It has been reported as cardiotonic, hepatoprotective, and anti-hyperlipidemic.
- **Use in alcoholism:** It reduces the oxidative stress [11].

2.3.6 BEARBERRY

- **Synonyms:** Arbutus Uva- Ursi, Uva- Ursi.
- **Biological source:** *Arbutus uva-ursi* belongs to the family *Ericaceae.*
- **Chemical constituents:** Uva-ursi includes gallic acid, ursone (resin substances). The bearberry leaves contain crystallizable glucoside as Arbutin, ericolin, and other phenolic constituents.
- **Pharmacological actions:** It is mostly used in urinary tract infection and inflammation.
- **Use in alcoholism:** The dose of 2-4 ml tincture (internal treatment) with 2-3 doses/day reduces alcohol craving over the time [12].

2.3.7 BRAHMI

- **Synonym:** Mansiktapatni (Hindi), Kapotavanka (Sanskrit), Janam, Nirali (Marathi).
- **Biological source:** Brahmi scientifically known as the *Bacopa monnieri* belongs to family *Scrophulariaceae.*
- **Chemical constituents:** It contains alkaloids (Brahmin), saponin and sterol. Later, numereous compound was isolated including nicotine, herpestine, betulic acid, stigmastarol, β-setosterol as well as bacoside (bacoside A and bacoside B), and nacopasaponin.
- **Pharmacological uses:** It is used as nontropic herb for centuries and also as antispasmodic, alterative, astringent, cardio tonic, diuretic, anticonvulsant, anti-inflammatory, analgesic, antipyretic, and antiepileptic.
- **Use in alcoholism:** Brahmi is a powerful herb to take during the treatment of addiction issues including drug, alcohol addiction. Its effect on the brain and the nervous system which reduces the cravings and calm the emotional eruption, occurs during the withdrawal process [13].

2.3.8 BITTER GOURD

- **Synonym:** Balsam pear (English), Karela (Hindi).
- **Biological source:** Bitter gourd obtained from plant *Momordica charantia* belongs to family *Cucurbitaceae.*
- **Chemical constituents:** It contains momordicoside and *p*-methoxy-benzoic acid.
- **Pharmacological use:** In treatment of neurodegenerative diseases, diabetes, inflammation, viral infections, autoimmune diseases, digestive system disorders, and also have antioxidant properties.
- **Use in alcoholism:** It is drink with buttermilk every morning to curb alcohol addiction. The juice obtained from the leaves of bitter gourd is a good cure for alcohol addiction [14].

2.3.9 CATNIP

- **Synonyms:** Catswort (English), catwort, catmint.
- **Biological source:** *Nepeta cataria* belong to family *Lamiaceae.*
- **Chemical constituents:** This includes tannins, volatile oil (α and β napetalactone), essential oil and flavonoids. The dose includes 4 g of dried herb, for sedation which is usually given as tea.
- **Pharmacological use:** Plant generally shows sedative action, and carminative.
- **Use in alcoholism:** It reduces the alcohol withdrawal symptoms [15].

2.3.10 CAYENNE PEPPER

- **Synonyms:** American pepper (English).
- **Biological source:** It is the dicotyledonous and short-lived perennial herb of species *Capsicum annuum* belonging to family *Solanaceae.*
- **Chemical constituents:** It contains 1.5% capsaicinoids, 0.11% capsaicin, 6,7- dihydrocapsaicin, nordihydrocapsaicin, homodihydro-capsaicin, homocapsaicin, and fixed oils.
- **Pharmacological use:** It is used as digestive stimulant.
- **Use in alcoholism:** Addition of Cayenne pepper to foods helps to reduce the cravings for alcohol and help to increase appetite. Since one of the primary concerns related to alcohol withdrawal is the reduced appetite and nausea [16].

2.3.11 CHAMOMILE

- **Synonyms:** Camomile (English), German chamomile, Flos chamomile.
- **Biological source:** *Matricaria chamomilla* is a well-known medicinal plant species from the family *Asteraceae.*
- **Chemical constituents:** Herniarin and umbelliferone (coumarin), chlorogenic acid and caffeic acid (phenylpropanoids), apigenin, apigenin-7-O-glucoside, luteolin and luteolin-7-Oglucoside (flavones), quercetin and rutin (flavonols), and naringenin (flavanone) are found in chamomile extract.
- **Pharmacological use:** It is used in hay fever, inflammation, muscle spasms, insomnia, sedative, and relaxant.
- **Use in alcoholism:** Chamomile tea is extremely beneficial for those who want to fend off alcohol cravings. Chamomile tea is especially useful for reducing alcohol cravings because it eliminates the restlessness for liquor [17].

2.3.12 EVENING PRIMROSE OIL

- **Synonyms:** Oenothera, Evening primrose (English).
- **Biological source:** It is a plant of species *Oenothera biennis* belonging to family *Onagraceae.*
- **Chemical constituents:** It contains cis-linoleic acid, cis-GLA, cis-6,9,12-octadecatrienoic acid, small amount of oleic, palmitic, stearic acids, steroids, and beta-sitosterol.
- **Pharmacological use:** It is effective for treating rheumatoid arthritis and diabetic neuropathy.
- **Use in alcoholism:** GLA from Evening Primrose oil prompts the brain to produce a specific type of prostaglandin called prostaglandin E, which indirectly protects the stomach lining, liver, and nervous system, and therefore protects the harmful local effects on the organ. It also fights the alcohol withdrawal symptoms [18].

2.3.13 GILOY

- **Synonyms:** Giloe (English), Amrita (Sanskrit), Gulvel (Hindi).
- **Biological source:** *Tinospora cordifolia* belongs to family *Menispermaceae.*

- **Chemical constituents:** Alkaloids, diterpenoid, lactones, glycosides, sesquiterpenoid, steroids, polysaccharides, etc. are the chief phytochemicals.
- **Pharmacological use:** Guluchi increases the memory power, enhances natural immunity in the human body, and having anti-oxidant nature.
- **Use in alcoholism:** Tinospora cordifolia extract proved its hepatoprotective effect against alcoholism not only by lowering the levels of these detected FAAs (fatty acid amides), but also by decreasing the level of liver-specific enzymes and lipids [19].

2.3.14 GINGKO

- **Synonyms:** Gingko (English), Maidehair tree, Ginkgo biloba.
- **Biological source:** *Ginkgo biloba* belongs to family *Ginkgoaceae.*
- **Chemical constituents:** This includes as flavonoids, (ginkgolide), and other constituent like quercitin, and kaempferol, terpenoids, resins, tannins, fatty acid, and flavones (ginkgolic acid).
- **Pharmacological use:** Ginkgo herb generally used in depression condition.
- **Use in alcoholism:** Ginkgo may be helpful in the treatment of alcoholism or in decreasing the alcohol consumption [20].

2.3.15 GINSENG

- **Synonyms:** Radix ginseng (English), Shin-seng, Panax root, Ginseng root.
- **Biological source:** It is the dried root of *Panax ginseng* belonging to family *Araliaceae.*
- **Chemical constituents:** It contains ginseng saponins, ginseng oils, phytosterols, carbohydrates, organic acids, peptides, vitamins, and minerals.
- **Pharmacological use:** It has antioxidant and anti-inflammatory properties.
- **Use in alcoholism:** Ginseng naturally in the system breaks and flushes away the toxins from the alcohol that linger in the body through long-term alcohol use. In the stomach and digestive tract it also lowers down the rate at which alcohol is absorbed. It reduces the symptoms of withdrawal but is also a way to manage cravings [21].

2.3.16 GOLDEN SEAL

- **Synonyms:** Indian Turmeric (English), Orange Root.
- **Biological source:** It is a perennial herb of species *Hydrastic canadensis* belonging to the family *Ranunculaceae.*
- **Chemical constituents:** It contains isoquinolone alkaloids such as hydrastine (1.5-4%), berberine (2.5%), canadine (0.5%), and other alkaloids.
- **Pharmacological use:** Golden seal has anti-viral and anti-fungal properties.
- **Use in alcoholism:** This herb can be a very effective remedy for stopping the desire for alcohol by selectively eradicating the intestinal factors involved in alcohol cravings [22].

2.3.17 GULBANAFSHA

- **Synonyms:** Wood violet (English), Banafsa (Hindi), Banafsha (Sanskrit).
- **Biological source:** *Viola odorata* belongs to family Violaceae.
- **Chemical constituents:** This includes rich source of sodium, calcium, iron, and magnesium. It also contain triterpene ketone, friedelin (0.016%), and β-sitosterol (0.033%) The dose is recommended as root infusion in dose 15-30 ml twice a day, 5-6 leaves can be chewed. Flowers are taken in dose half to five grams generally flowers are mostly used.
- **Pharmacological use:** It is used in treating the respiratory disorders and insomnia conditions.
- **Use in alcoholism:** It is also used to relief alcohol-induced stress [23].

2.3.18. GREEN TEA

- **Synonyms:** Camellia.
- **Biological source:** It is obtained from the tea plant *Camellia sinensis* belongs to the family *Theaceae.*
- **Chemical constituents:** It contains catechin, epicatechin, epigallocatechin, epigallocatechin gallate, proteins, chlorophyll, and amino acids.

- **Pharmacological use:** It shows anti-aging, anti-Alzheimer, anti-parkinsonism, anti-stroke properties, and cardiovascular diseases.
- **Use in alcoholism:** It is rich in tannins, mainly catechins as epicatechin, epigallocatechin, etc. which helps in the detoxification of body due to consumption of alcohol [24].

2.3.19. HOVENIA DULCIS

- **Synonyms:** Japanese raisin tree.
- **Biological source***:* *Hovenia dulcis* was obtained from plant known as Japanese raisin tree belongs to the family *Rhamnaceae*.
- **Chemical constituents:** It contains kaemperol, myricitin, dihydromyricetin, chrysophanol, ethyl caffeate, 3-hydroxy-4-methoxybenzoic acid, β-sitosterol, epicatechin, and vanillin.
- **Pharmacological use:** It shows hepatoprotective effect, anti-diabetic effect, antimicrobial, and anti-parasitic effect.
- **Use in alcoholism:** The herb *H. dulcis* was found to effective in alleviating intoxication. It contains the derivative known as dihydromyricetin (DHM). This DHM containing *H. dulcis* has the potential to treat alcohol withdrawal symptoms [25].

2.3.20 LIQUORICE

- **Synonyms:** Glycyrrhiza (English), Mulethi (Hindi), Jethi madh (Sanskrit), Yashtimadhu (Bengali).
- **Biological source:** It is a peeled and unpeeled stolons, roots and subterranean stems of *Glycyrrhiza glabra* belonging to family *Leguminosae*.
- **Chemical constituents:** It contains glycyrrhizin, liquirtin, isoliquertin, liquiritigenin, rhamnoliquirilin, and five new flavonoids; glucoliquiritin apioside, prenyllicoflavone A, shinflavanone, shinpterocarpin, and 1-metho-xyphaseolin.
- **Pharmacological use:** It is anti-inflammatory, expectorant, controls coughing, asthma, peptic ulcer, herpes, etc.
- **Use in alcoholism:** Liquorice root is a safe and effective remedy for safely quitting alcohol as this plant has strong anti-inflammatory and antioxidant effects by enhancing the body's natural barrier against pathogens and strengthening the immune system [26].

2.3.21 MILK THISTLE

- **Synonyms:** Marian thistle, Mediterranean milk thistle, and Scotch thistle (English) Ksheerivriksha (Sanskrit), Paal Chedy (Malayalam).
- **Biological source:** *Silybum mariamum* belongs to the family *Asteraceae*.
- **Chemical constituents:** The plant contains at least seven flavolignans and the flavonoid taxifolin. The most important flavolignans present include silybin, silydianin, and silychristine. Silybin represents between 50% and 70% of the extract from silymarin.
- **Pharmacological use:** It is help to is the symptoms of irritable bowel syndrome and mild laxative, mild anti-inflammatory and anti-bacterial effect.
- **Use in alcoholism:** Silymarin act as hepatoprotectant against toxins occur due to the over-consumption of alcohol. Milk thistle is effective for treating and preventing alcohol-liver diseases. They were also effective in individual who begins to experience early sign of alcoholic liver disease [27].

2.3.22 NAGKESHAR

- **Synonyms:** Mesua, Naghas (Hindi), Ahipushpa (Sanskrit).
- **Biological source:** *Mesua ferrea* belongs to the family *Guttiferae*.
- **Chemical constituents:** This includes mensuol, sitosterol, fatty acids, mesuanic acid, and a bioflavonone called as mesuaferrone A.
- **Pharmacological use:** The plant generally shows anti-convulsant and anti-inflammatory property. The dose includes two capsules twice daily, with plain water after meals, and powder 1–3 g.
- **Use in alcoholism:** It reduce the alcohol withdrawal and rewarding symptoms [28].

2.3.23 NISOTH

- **Synonyms:** Trivrit (Sanskrit), Pithori (Hindi).
- **Biological source:** *Ipomoea Turpethu* belongs to family Convolvulaceae.

- **Chemical constituents:** This includes resin glycosides, saponins, flavanoids, steroids, carbohydrates, starch, volatile oil, scopoleptin, sitosterol, glucose, and rhamnose.
- **Pharmacological use:** It helps in easy purgation, losing weight, ascites, and leukoderma. The use of root powder gives relief in habitual constipation.
- **Use in alcoholism:** Anti-alcoholic property of herb inhibits and induces aversive action to alcohol consumption they helps patients reduce alcohol intake [29].

2.3.24 PASSION FLOWER

- **Synonyms:** Passiflora.
- **Biological source:** It is a flowering plant of species *Passiflora incarnate* belonging to family *Passifloraceae.*
- **Chemical constituents:** It contains quercetol, kaempferol, apigenin, luteolin, vitexin, saponaroside, schaftoside, isoschaftoside, sitosterol, and stigmasterol.
- **Pharmacological uses:** It is used in hypertension, tachycardia, palpitations, migraines, dizziness, and ulcers.
- **Use in alcoholism:** The plant was found to have great anti-oxidant potential which effectively reduces the stress produced by consumption of alcohol [30].

2.3.25 SKULLCAP

- **Synonyms:** Scutellaria.
- **Biological source:** It is a flowering plant of species *Scutellaria pekinensis* belonging to family *Lamiaceae.*
- **Chemical constituents:** It contains baicalein, baicalin, apigenin, oroxylin A, scutellarein, and skullcapflavone.
- **Pharmacological uses:** It is used as a natural tranquilizer. It is a great remedy for stress and anxiety. Its ability of increasing oxygen (via blood flow) to the brain may also help in preventing strokes as well.
- **Use in alcoholism:** This herb is very useful for treating withdrawal from drugs and alcohol [31].

2.3.26 ST. JOHN'S WORT

- **Synonyms:** Amber, Balsana, Demon chaser, Hypericum.
- **Biological source:** *Hypericum perforatum* belongs to family *Hypericaceae*.
- **Chemical constituents:** This includes, main constituent such as hypericin, pseudohypericin, chlorogenic acid, rutin, resin, tannins, apigenin-7-O-glucoside, quercetin, and vitexin.
- **Pharmacological uses:** It is used in the treatment of anxiety, mild to moderate depression, and in insomnia.
- **Use in alcoholism:** The extract of St. John's Wort suppresses alcohol intake in animal models of excessive drinking with minimal effects on other appetitive behaviors [32].

2.3.27 THUMBERGIA LAURIFOLIA

- **Synonym:** Neel lata (Hindi).
- **Biological source:** *Thunbergia laurifolia* is a popular ornamental vine in the tropics belongs to family *Acanthaceae*.
- **Chemical constituents:** Compounds isolated from the leaves included iridoid glucosides, grandifloric acid, glucopyranosides, and derivatives of apigenin.
- **Pharmacological use:** They were used as an antidote for poisons and drugs, including treatment of drug addiction.
- **Use in alcoholism**: It shows alcohol detoxifying activity in liver [33].

2.3.28 VIDARIKAND

- **Synonyms:** Swadukanda (Sanskrit), Vidarikand (Hindi).
- **Biological source:** *Pueraria tuberose* belongs to the family Leguminosae.
- **Chemical constituent:** It includes carbohydrates content (64%), and proteins (10.9%). The recommended doses tuber powder is 3-6 g with milk.
- **Pharmacological uses:** They mainly show the action on urinary systems, and have good expectorant property.
- **Use in alcoholism:** It reduces the alcohol withdrawal symptoms [34].

2.4 CONCLUSION

From the above literature review, it was found that alcoholism is a serious problem throughout the world. It hampers the balance of body and creates dependency for it. The above information has included the botanical characteristics of the plant which helps in identification of the plant, ethnobotany which give traditional use of the plant, and the reported activities of the plant. Various synthetic drugs are available for the treatment of alcoholism but they were severe side effects and contraindication regarding the use of these drugs. Nowadays world is getting attention for natural remedies. The therapeutic effects of these natural remedies may be less pronounced at times than synthetic drugs, but the probability of adverse symptoms to minimum. By this information, it can be concluded that in the core of the nature there are so many plants which possess antialcoholic property. Herbal medications are free from side effects and toxicity unlike the synthetic ones. So theses natural actives can be easily and safely replaced as antialcoholics as a part of treatment.

KEYWORDS

- **addiction**
- **alcohol intake**
- **alcoholism**
- **anti-alcoholism**
- **ethnopharmacology**
- **natural remedies**

REFERENCES

1. World Health Organization (WHO), (2004). *WHO Global Status Report on Alcohol 2004*. Geneva: WHO.
2. David, A. F., Carrington, R., & Patrick, G. O., (2000). New therapies for alcohol problems: Application to primary care. *Am. J. Med.*, 227–237.
3. Oscar-Berman, M., Barbara, S., Denise, L. E., & Charles, E., (1997). Impairment of brain and behavior: The neurological effects of behavior. *Alcohol Health and Research World, 21*(1), 65–75.
4. Ernst, D. B., Pettinati, H. M., Weiss, R. D., et al., (2008). An intervention for treating alcohol dependence: Relating elements of medical management to patient outcomes with implications for primary care. *Ann. Fam. Med., 6*, 435.

5. Concept & Definition. https://www.alcoholwebindia.in (accessed on 15 July 2020).
6. Goh, E. T., & Morgan, M. Y., (2017). Review article: Pharmacotherapy for alcohol dependence-the why, the what and the wherefore. *Alimentary Pharmacology and AMP; Therapeutics*, 45(7).
7. Saka, W. A., Akhigbe, R. E., Ishola, O. S., Ashamu, E. A., Olayemi, O. T., & Adeleke, G. E., (2011). Hepatotherapeutic effect of aloe vera in alcohol-induced hepatic damage. *Pakistan Journal of Biological Sciences, 14*(14), 742–746.
8. Chen, K. H., Lin, B. R., Chien, C. T., et al., (2011). Emblica officinalis gaertn. attentuates N-nitrosodiethylamine-induced apoptosis, autophagy, and inflammation in rat livers. *J. Med. Food, 14*, 746–755.
9. Home remedies for alcoholism cited from http://www.searchhomeremedy.com/herbal-remedies-for-alcohol-cravings/ (accessed on 15 July 2020).
10. www.findhomeremedy.com/natural-treatment-for-alcoholism (accessed on 15 July 2020).
11. Amalraj, A., Gopi, S., (2017). Medicinal properties of Terminalia arjuna (Roxb.) Wight & Arn.: a review. *Journal of traditional and complementary medicine. 7*(1), 65-78..
12. Cynthia Wright. Arctostaphylos Uva-Ursi, Bearberry. http://www.npshistory.com/publications/lewi/bearberry.pdf
13. (accessed on 15 July 2020).
14. Mathur, D., Goyal, K., Koul, V., & Anand, A., (2016). The molecular links of re-emerging therapy: a review of evidence of Brahmi (Bacopa monniera). *Frontiers in pharmacology. 7*, 44.
15. Madhu, G., Sushil, S., Ajay, K. G., & Rekha, B., (2011). *Momordica Charantia Linn.* (Karela): Nature's silent healer. *International Journal of Pharmaceutical Sciences Review and Research*, 11(1), 32–37, Article-007.
16. Manisha, S., Rajat, R., & Vikramaditya, P. N. V., (1995). Pharmacognosy of *Nepeta cataria. Ancient Science of Life, 14*, 225–234.
17. Xu BJ., Zheng Y. N., & Sung C. K., (2005). Natural medicines for alcoholism treatment: a review. *Drug and alcohol Review. 24*(6), 525-36.
18. Ompal, S., Zakia, K., & Manoj, K. S., (2011). Chamomile (*Matricaria chamomilla* L.): An overview. *Pharmacognosy Reviews* (Vol. 5, No. 9, pp. 82–95), Wolters Kluwer, Medknow Publication.
19. www.herbwisdom.com/herb-evening-primrose.html (accessed on 15 July 2020).
20. Soham, S., & Shyamasree, G., (2012). *Tinospora cordifolia*: One plant, many roles. *Ancient Science of Life, PMC*, (Vol. 31, No. 4, pp. 151–159) Wolters Kluwer, Medknow Publication.
21. Harris, R. L., Mark, D. K., Victor, L. F., & Sanjiv, A., (2017). Moderate doses and commercial preparation of *Ginkgo biloba* do not alter markers of liver function but moderate alcohol intake does: A new approach to identify and quantify biomarkers of "adverse effects" of dietary supplements. *Science Direct, 84*, 45–53.
22. www.the-alcoholism-guide.org/herbal-treatment-for-alcoholism.html (accessed on 15 July 2020).
23. www.howtostopdrinkingalcohol.net/how-to-give-up-drinking-alcohol (accessed on 15 July 2020).
24. Feyzabadi, Z., Ghorbani, F., Vazani, Y., & Zarshenas, M. M., (2017). A critical review on phytochemistry, pharmacology of Viola odorata L. and related multipotential products in traditional Persian medicine. *Phytotherapy research.31*(11) 1669-75.

25. Dorchies, O. M., Wagner, S., Waldhauser, K. M., et al., (2003). Anti-fibrotic properties of green tea catechins on mouse muscle cell cultures. *Neuromuscul Disord., 613*, 639.
26. Quang, U. L., Horng, L. L., Ming, C. W., Rakesh, K. J., Duy, L. N., & Thi, H. H. N., (2018). *Hovenia dulcis* thunb. *Journal of Pharmacognosy and Phytochemistry, 7*(6), 1137–1141.
27. Carai, M. A., Agabio, R., Bombardelli, E., et. al., (2000). Potential use of medicinal plants in the treatment of alcoholism. *Fitoterapia. 71* S38-42.
28. Iman, N., Ebrahim, T., & Zahra, N., (2016). Study on *Silybum marianum* seed through fatty acids comparison, peroxide tests, refractive index and oil percentage. *Pharmacognosy Journal, 8*(6), 595–597.
29. Chahar, M. K., DS SK., Geetha, L., et. al., (2013). A review of the medical evidence for its phytochemistry and pharmacological actions. *African Journal of Pharmacy and Pharmacology, 7*(6), 211-9.
30. Tasleem, A., Mohd, K. H., Mohd, T., Javed, I. S., Mohd, K., Mohammed, W. A., & Munawwar, H. K., (2017). A Review on *Operculina turpethun*: A potent herb of unani system of medicine. *Journal of Pharmacy and Phytochemistry, 6*(1), 23–26.
31. www.healthline.com/health/anxiety/calming-effects-of-passionflower#3 (accessed on 15 July 2020).
32. Herbs list from www.herbslist.net/skullcap.html (accessed on 15 July 2020).
33. Sweety, M., (2012). *Pharmacognosy of St. John's Wort.* Pharma Xchange, https://pharmaxchange.info/2012/12/pharmacognosy-of-st-johns-wort/ (accessed on 15 July 2020).
34. Eric, W. C. C., Suit, Y. E., Yuen, P. T., & Zhiew, C. W., (2011). Phytochemistry and pharmacological properties of *Thunbergia laurifolia:* A Review. *Pharmacognosy Journal, 3*(24), 1–6.
35. Maji, A. K., Pandit, S., Banerji, P., Banerjee, D., & Pueraria tuberose., (2014). A review on its phytochemical and therapeutic potential. *Natural product research, 28*(23), 2111-27.

CHAPTER 3

Reviewing the Aphrodisiac Activity of *Lepedium meyenii* Plant in Modern Contexts

SHILPA S. BORKAR,[1] DEBARSHI KAR MAHAPATRA,[2] and SEEMA USARE[1]

[1]Department of Pharmacology, Kamla Nehru College of Pharmacy, Nagpur – 441108, Maharashtra, India

[2]Assistant Professor, Department of Pharmaceutical Chemistry, Dadasaheb Balpande College of Pharmacy, Nagpur – 440037, Maharashtra, India

ABSTRACT

Lepidium meyenii, usually called Maca, maybe a Peruvian plant that grows high within the mountain chain, in areas over four, 4000 meters on top of water level. It contains most of the essential amino acids and twenty of the essential fatty acids, further as several vitamins, minerals, and several other sterols and glucosinolates required by man. The clinical studies of varied crude extracts from differing kinds of maca have multiple applications such as nutritionary, chemoprevention, aphrodisiac, and fertility-enhancing functions. The particular and distinctive unsaturated acids and amides found solely during this plant are the macaenes and macamides contained at intervals the hypocotyls.

3.1 INTRODUCTION

Maca, conjointly called *Lepidium meyenii* Walpers, maybe a Peruvian plant growing in Peruvian Central Andes in areas on top of four, 4000 meters belonging to the family Brassicaceae and 150–175 species of the Lepidium are wide on the market. German phytologist Gerhard Walpers in 1843 first described this plant and cultivated it within the chain mountain at altitudes from 3500 to 4450 m on top of water level in the puna and suni surroundings. Maca is that the sole species cultivated as a starch crop and it contains several proteins, amino acids, fats, and vitamins, also as a range of secondary metabolites like macaene, alkaloids (including maca amide), glucosinolates, and alternative elements [1].

It is conjointly an upscale supply of micro- and macroelements, together with iron, calcium, copper. Maca has been utilized by native individuals as a food ingredient and in ancient people drugs to boost sexual performance, fertility, energy, alertness, mental concentration, mood, and physical immunity and for improvement of memory and anti-viral activity and inhibitor properties [2, 3]. A comparison of designated studies is shown in Table 3.1.

3.2 BACKGROUND CLASSIFICATION AND CULTIVATION

L. meyenii may be dried stalk root happiness to the mustard family [4, 5]. Chacon modified its name to *L. peruvianum*, as a result of herbarium specimens from Bolivia and Argentina were classified as *L. meyenii* [6]. It is been urged that the cultivated maca of these days is not *L. meyenii*, however, the new species *L. peruvianum* whereas most of the maca sold-out in commerce remains are stated as *L. meyenii*. *L. meyenii* propagated by seed, need 7–9 months to supply the harvested root. It cultivated on rocky soil on rough Andean terrain temperatures between –20°C and 20°C. The soil used is acidic clay or stone having a ratio of roughly seventieth. Maca is usually planted from September to October at the start of the time of year, and harvest it from could to July when a vegetative part of 260–280 days.

3.3 TRADITIONAL AND MEDICATIVE USES

Maca was cultivated by Pumpush, Yaros, and Ayarmaca Indians for food. For the treatment of fertility issues at high altitudes, conquistadors fed the baked or cooked root powder is given to animals and also the Chinchaycoch as Indians used it in bartering. In South America, the sweet aromatic dish of dried maca is consumed underneath the name mazamorra. In Huancayo, Peru, maca jam and pudding are well-liked, and maca is usually created into a sweet, fragrant, soured drink known as maca stylish. It is conjointly

TABLE 3.1 Comparison of Selected Studies

Type of Maca	Plant Part	Origin	Type of extract	Study subjects	Results
Yellow	Dried hypocotyls	Pasco, Peru	Mixed in water Water boiled extract	Rats	Dermal protection against UV radiation was shown; the most effective-boiled maca
Yellow, Red Black	Dried hypocotyls	Carhuamayo, Junin, Peru	Water boiled extract	Rats	No effect Increase in sperm mobility
Black	Dried hypocotyls	Carhuamayo, Junin, Peru	Hydro-alcoholic extract	Mice	Inhibition of memory disorder
No data	Powdered root	Pampas Valley (Tayacaja, Huancavelica)	Methanol extract	Cell lines	Inhibition of influenza viruses A and B
Black	Dried hypocotyls	Carhuamayo, Junin, Peru	Water boiled extract, frozen and lyophilized	Mice	Improved learning and memory in mice
Black, Yellow, Violet	Dried plant	Juhua Village market Kunming province, Yunnan	Petroleum ether extract	Rats	Strong antioxidant effects (purple Maca) in rats
Red, Yellow, Black	Dried hypocotyls	Carhuamayo, Junin, Peru	Water-boiled extract	Rats	Increases in sperm cell count, increases in body weight in animal models
No data	Dried root	Kunming, China	Polysaccharide mp21	Cell lines	Immuno-stimulating effects, increased macrophage activity
No data	Dried hypocotyls	Arequipa, Peru	Methanol extract	Cell lines	Neuroprotective effects
	Leaves	Ramancancha, province of Junin		Rats	

TABLE 3.1 *(Continued)*

Type of Maca	Plant Part	Origin	Type of extract	Study subjects	Results
No data	No data	Peru	No data	Mice	Regulation of autophagy proteins, a slowdown of cognitive decline associated with age
Yellow	No data	Ecoandino SAC Co. Lima, Peru	Lipid extract	Rats	Improvement of physical condition by suppression of oxidative stress
No data	No data	Quimica Suiza, Peru	Powder	Rats	Reduction of plasma cholesterol, VLDL, LDL, and TAG
No data	Root	Junin, Peru	Powder	Women	Suppression of menopausal symptoms
No data	Root	Lima, Peru	Powder	Men	No effect on the level of sex hormones
No data	No data	Linzhi, Xizang, China	Petroleum ether extract	Mice	The antioxidant effect in brain tissue
Black, Red, Yellow	Dried bulb	MACAXS™ (TOWA Corporation K.K., Tokyo, Japan)	Hydro-alcoholic powder extract	Rats	Stimulation of male sex hormones

supposed to control secretion, stimulate metabolism, and improve memory, and is touted for medicinal drug and antitumor properties, moreover as for set anemia, leukemia, and AIDS. Because of its wide spectrum of reputed qualities, maca is additionally referred to as Peruvian ginseng [7]. The dried root contains 13–16% macromolecule and is made in essential amino acids, whereas the recent root is high in iodine and iron. It contains 250 mg of Ca, a pair of grams of K, and 15 mg of iron in a hundred grams of dried root, and sterols (0.05–0.1%), minerals, and vitamins. Maca contains 3.72% fatty acids, as well as saturated fatty acid, dodecanoic acid, tetradecanoic acid, hexadecanoic acid, palmitoleic acid, octadecanoic acid, oleic acid, linolic acid, and omega-6 [10]. Additionally, a replacement acyclic keto acid, 5-oxo-6*E*,8*E*-octadecadienoic acid has been isolated from the tubers [11]. Yellow maca is found to possess higher macromolecule and macromolecule content compared to the red and black varieties [9].

3.4 SECONDARY METABOLITES

The major secondary metabolites gift in *L. meyenii* are often classified into four groups: (a) essential oils; (b) glucosinolates; (c) alkaloids; and (d) macamides. The volatile oil of maca contains a minimum of 53 elements [12]. The foremost elements of the steam distilled oil are phenyl acetonitrile (85.9%), benzaldehyde (3.1%), and 3 methoxyphenylacetonitrile (2.1%). Maca contains two categories of unsaturated fatty acids, the macaenes, and macamides [13]. The macamides, additionally referred to as radical alkamides, are secondary metabolites that are found solely in genus *Lepidium meyenii* [14, 15]. The most macamides are known as *N*-benzylhexadecanamide, *N*-benzyl-(9Z)-octadecenamide, *N*-benzyl-(9*Z*,12*Z*)-octadecadienamide, *N*-benzyl-(9*Z*, 12*Z*, 15*Z*)-octadecatrienamide, and *N*-benzyloctadecanamide. Total macamides in dried stuff has been found to vary from 0.0016–0.0123% [16] (Table 3.2).

3.5 PRINCIPAL APHRODISIAC MECHANISM

Infertility could be a biological, clinical, personal, and social downside that the malefactors contribute [22]. The term "Aphrodisiac" has been derived from Greek, an image of affection, and sweetness in classical mythology. Ancient literature and records show the deep interest of persons for substances to extend sexual desire, efficiency and pleasure because it takes

TABLE 3.2 Identification of the Main *Lepidium meyenii* (Maca) Metabolites by LC-MS-DAD and -qTOF Analyses Operating in Positive Ion Mode (ESI +)

Name	RT (min)	ESI(+)/ Expected (m/z)	ESI(+)/ Measured (m/z)	Delta (ppm)	Main Fragment (m/z)	λ max (nm)
(1R,3S)-1-methyltetrahydro-beta-5,6-hydridecarboline-3-carboxylic acid	13.842	233.1284	233.1276	3.431585341	91.054	215.5
N-(3-hydroxy-benzyl)-2Z-fivecarbon acrylamide	18.693	180.1019	180.1019	0.027	162.0914	220.1
1-dibenzyl-2-propane-4,5-dimethylimidazilium	18.997	227.1542	227.1546	1.76091835	91.0544	219.5
3-benzyl-1,2-dihydro-N-hydroxypyridine-4-carbaldehyde	22.571	216.1175	216.1173	0.916936972	91.0543	213.2
1,3-dibenzyl-2,4,5-trimethylimidazilium	24.574	291.1854	291.1851	1.030271435	91.054	214.1
1,3-dibenzyl-2-phenyl-4,5-dimethylimidazilium	27.749	353.2014	353.2013	0.283124586	185.1072	221.1
N-ethyl-tetradecene ester	28.187	274.274	274.2743	1.09379671	256.2637	221.2
1,3-dibenzyl-2-pentyl-4,5-dimethylimidazilium	29.578	347.2481	347.2482	0.28797854	277.1705	222.7
5-oxo-6E,8E-octadecadienoic acid	40.444	295.2267	295.2266	0.338722751	277.2165	223.9
N-benzyl-5-oxo-6E,8E-octadecadienamide	43.789	384.2824	384.2822	0.520450585	306.2421	224.3
N-benzyl-9-oxo-12E-octadecenamide	44.976	386.2981	386.2979	0.517734879	261.2212	223.8
N-(3-methoxybenzyl)-(9Z,12Z,15Z)-octadecatrienamide	48.505	398.298	398.2976	1.004273182	381.2785	223.5
N-benzyl-(9Z,12Z,15Z)-octadecatrienamide	48.863	368.2875	368.288	1.357635	351.2678	223.9
N-(3-Methoxybenzyl)-(9Z,12Z)-octadecadienamide	51.403	400.3137	400.3133	0.999216364	365.2838	224.3
N-Benzyl-(9Z,12Z)-octadecadienamide	51.868	370.3031	370.3034	0.81014715	232.1694	224.7
N-benzylhexadecanamide	54.34	346.3031	346.3034	0.86629314	221.2261	221.3
N-benzyl-9-Z-octadecenamide	55.169	372.3188	372.3194	1.61152217	247.2423	225.9
N-benzyloctadecanamide	59.184	374.3345	374.3344	0.267140752	296.3172	225.5

care of the disorders of desire, ejaculation, climax, and dysfunction [23]. In the Ayurvedic medicative system of India, a separate section is dedicated to curing dysfunction with the assistance of natural merchandise and its standard as "Vajikaran Medical Care" or Virlification. Despite the provision of standard artificial aphrodisiacs with their precise mechanism of action, many scientists worked to find out a natural substance of basic cognitive process that it may have less or no aspect effects [24]. Artificial aphrodisiacs like the anti-impotence drug, tadalfil, and verdenafil that work by PDE_5 inhibition may cure dysfunction and loss of libido; but, a natural aphrodisiac may have further advantages like providing nutrition and facilitate recover physiological state with none adverse effects [25].

3.6 BRAIN AND NEUROCHEMICAL BASIS OF SEXUAL BEHAVIOR

Drugs touching gender will either act on the Central Nervous System (Brain) and/or on the peripheral nervous system. Medicine touching the brain and presumptively sex centers are usually attributed to rise or decrease in arousal. Medicine that affects peripheral nerves won't affect arousal directly however, might affect sexual operation. In some cases, medicine action is direct and involves chemical alteration of the neurons, that governs arousal or operate or else, some medicine might act indirectly by fixing blood flow to the reproductive organ. Most hypotheses regarding the organic compound basis of sexual behavior are derived from studies in animals, however, in some cases support has been provided by clinical studies. Five major neurochemically distinct systems are alleged to work along for increasing arousal. The transmitters embrace vasoconstrictive, dopamine, serotonin, neurotransmitter, and amine [26]. The foremost wide supported hypotheses recommend that each monoamine neurotransmitter associate degreed monoamine neurotransmitter are concerned within the organic compound management of sexual behavior with serotonin taking part in a restrictive role and dopamine a stimulative role. Monoamine neurotransmitter plays a vital role in the central management of sexual behavior in males [27]. An increase within the activity of central dopaminergic systems correlates with sexual issues [28]. In vivo microdialysis in acutely aware male rats unconcealed that monoamine neurotransmitter transmission will increase sharply within the striate body, nucleus accumbens, and medial preoptic space throughout sexual relation [29–31]. This transformation in central neurotransmission could also be permissive to a series of motor responses together with erectile organ erection. It should additionally modulate the activity of brain nuclei directly concerned within the management of erectile organ erection [32]. As an example, medicine like dopamine, that increase levels of monoamine neurotransmitter within the brain, tend to be related to increasing sexual desire and increased sexual operate in patients tormented by abnormal monoamine neurotransmitter activity like that associated with Parkinson's malady. In distinction, medicine obstruction monoamine neurotransmitter operates like major tranquilizer cause loss of arousal. It is been long suspected that monoamines play a vital role within the regulation of sexual behavior, significantly that of dopaminergic transmission that is, helpful to masculine activity and each dopaminergic and adrenergic receptors are concerned. Yohimbine, bromocriptine, and reserpine are alpha-adrenergic receptor blocking agents whereas yohimbine, bromocriptine, amphetamine, and apomorphine all come with the neurotransmitter dopamine for binding to membrane sites [33]. Moreover, some studies have also prompt that the monoamine neurotransmitter unleash is additionally exaggerated throughout sexual issues within the paraventricular nucleus of the neural structure which during this neural structure nucleus dopamine facilitates erectile organ erection and sexual behavior by activating NO production in the cell bodies of hormone neurons dominant penile erection and sexual motivation, that project to extrahypothalamic brain areas and the funiculus [27, 34–38]. Therefore, there seem to be tons of cross speak at completely different somatic cell levels between monoamine neurotransmitter and element oxide; this has been mentioned more within the next section.

3.7 NITRIC OXIDE-BASED MECHANISM OF SEXUAL BEHAVIOR

Nitric oxide (NO) is a typical restrictive molecule having the twin role as a secondary messenger/neurotransmitter. It has been involved in numerous physiological functions [38]. Findings up to now indicate that NO may additionally be a significant somatic cell traveler [39]. Specifically, it is a longtime physiological negotiator of erectile organ erection [40] and within the brain; NO synthase is very focused in structures directly or indirectly concerned in sexual behavior (olfactory bulb, supraoptic and paraventricular nuclei, amygdala, body part structures, etc.) [41]. Recent studies recommend that NO may be a major physiological stimulant for relaxation of erectile organ vasculature and trabeculate swish muscle, essential for erectile organ erection [43]. Relaxation of the trabeculate swish muscle of the corpus cavernosa ends up in an attenuated tube-shaped structure resistance and exaggerated blood flow to the erectile organ. Aboard the exaggerated flow, blood vessel outflow is reduced by the compression of the subtunical venules. The mix of exaggerated influx and attenuated outflow causes erectile organ engorgement and erection. NO from the tube-shaped structure epithelial tissue of the sinusoids and from the nonadrenergic, noncholinergic, and cavernosal nerves seems to mediate the vasodilatation [43, 44]. The new drug used for the treatment of impotence and anti-impotence drug acts by potentiating the result of NO by inhibiting the precise catalyst phosphodiesterase-V that terminates the action of NO generated cGMP within the erectile organ vasculature [45]. Several healthful herbs and medicines derived from these herbs are shown to own effects on the NO signal pathway. As an example, the saponins from ginseng (ginsenosides) are shown to relax blood vessels (probably contributive to the anti-fatigue and blood-pressure-lowering effects of ginseng) and corpus cavernosum (thus, for the treatment of men tormented by erectile dysfunction; but, the legendary aphrodisiac result of ginseng could also be associate degree overstatement) [46].

3.8 ANDROGEN-BASED MECHANISM OF SEXUAL BEHAVIOR

Androgens play a vital role within the development of secondary male sexual organs like the duct, ductus deferens, cyst, prostate, and therefore, the erectile organ. Moreover, androgens are required for pubescence, male fertility, and male sexual operate [47]. An androgenic hormone is that the principal androgen secreted by the testes. An androgenic hormone is synthesized within the Leydig cells of the testes, aroused by gonadotrophic hormone (LH). One in all the principal effects of the androgenic hormone inside the testes is that the stimulation of gametogenesis in liquid body substance tubules. The testosterone- or dihydrotestosterone-receptor complicated next crosses the nuclear membrane, binds to deoxyribonucleic acid and stimulates new template RNA synthesis, and, thereby, new supermolecule synthesis. The result of the androgenic hormone on sexual desire might need the conversion of testosterone to estrogen within the neural structure. The mechanisms whereby androgenic hormone affects muscle, bone, and therefore, the erythron does not seem to need previous molecular conversion [48]. Medicine accustomed to treat numerous sexual issues is found to change the action of neurotransmitters that might be helpful, inhibitory, or both. Androgens are acknowledged to influence NO production within the brain furthermore as in the fringe [49, 50]. NO is synthesized by the catalyst gas synthase (NOS) that plays a vital role in several brain functions. NO operates as a neurochemical and NOS is a gift within the regions of the brain that regulate sexual functions [51]. Curiously, the administration of an androgenic hormone to gelded male rats will increase the amount of NO synthase labeled neurons within the mPOA, indicating a rise in NO synthesis [52]. NO is capable of stimulating monoamine neurotransmitter (DA) unleash within the mPOA, that successively stimulates erectile organ erection. This mechanism might represent a method during which androgens stimulate arousal [53].

3.9 MECHANISM OF *LEPIDIUM MEYENII*

The mechanism by which it plays its role as an aphrodisiac has been subjecting to exploration for several investigators. A study is incontestable that Maca extract does not move with the steroid receptor and it is no direct steroid hormone activity. The hypothesis that Maca acts by modulation of the hypothalamic-pituitary axis through regulation of secretion and balance was additionally found to be not line up to a giant extent as Maca was found to possess no impact on endocrine gland hormone levels; *viz.* FSH, LH, androgen, estrogen, and gonadotropin. Improvement of larginine-nitricoxide activity has also been attributed to Maca [54].

3.10 CONCLUSION

Male sexual dysfunction and connected sexual debilities arise as results of varied physiological, psychological, and clinical factors. Aphrodisiac plants are in demand to cure dysfunction and have the promise to switch terribly pricey and fewer safe artificial aphrodisiacs. The foremost necessary issue at the present is the investigation of a lot of aphrodisiac plants followed by isolation and identification of their active constituents. A lot of precise and properly orientating animal and clinical trials are required to judge aphrodisiac activity connected parameters besides safety studies. A more robust approach with inclusion of material medica, pharmacology, and pharmacodynamics is needed to avoid any conflict concerning activity and effectualness of those resources. any analysis of active constituents, doable modes of action, and pathways concerned may facilitate the United States to establish promising natural aphrodisiacs that might be a milestone within the management of male sexual dysfunction. The herbs can be an effective aphrodisiac, moreover, isolation, and identification of active constituents from plants may bring a dynamic change in the modern world.

KEYWORDS

- **aphrodisiac**
- **ethnopharmacology**
- ***Lepidium meyenii***
- **maca**
- **medicine**
- **phytochemicals**
- **traditional**

REFERENCES

1. Dorota, K., Barbara, J. J., Katarzyna, B., & Agnieszka, W. *Lepidium meyenii* (Maca)–multidirectional health effects – review. (2018). *Curr. Issues Pharm. Med. Sci*, *31*(3), 107–112.
2. Gerbarg, P. L., & Brown, R. P. (2013). Phytomedicines for prevention and treatment of mental health disorders. *Psychiatr Clin North Am*, *36*, 37–47.
3. Gonzales, G. F., Villaorduña, L., Gasco, M., Rubio, J., & Gonzales, C. (2014). Maca (*Lepidium meyenii* walp), a review of its biological properties. *Rev Peru Med Exp Salud Publica*, *31*(1), 100–110.
4. Toledo, J., et al.,. (1998). Genetic Variability of *Lepidium meyenii* and other Andean Lepidium Species (Brassicaceae) Assessed by Molecular Markers. *Ann Bot*, *82*(4), 523–530.
5. Quiros, C. F., & Cardenas, R. A. (1997). Maca (*Lepidium meyenii* Walp.). In M. Hermann., & J. Heller (Eds.), *Andean Roots and Tubers: Ahipa, Arracacha, Maca and Yacon* (pp. 173–198). Rome: International Plant Genetic Resources Institute.
6. Chacon, G. (1990). La maca (*Lepidiumper uvianum* Chacon sp. nov.) y su habitat. *Rev. Peru. Biol*, *3*, 171–272.
7. Maca. (2001). In: Der Marderosian, A. (Ed.), *Guide to Popular Natural Products*. St. Louis: Wolters Kluwer Company.
8. Dini, A., et al. (1994). Chemical composition of *Lepidium meyenii*. *Food Chemistry*, *49*(4), 347–349.
9. Comas, M., et al. (1997). Bromatological studies on *Lepidium meyenii*. *Alimentaria (Madrid)*, 286, 85–90.

10. Zheng, B. L., et al. (2000). Effect of a lipidic extract from *Lepidium meyenii* on sexual behavior in mice and rats. *Urology*, *55*(4), 598–602.
11. Muhammad, I., et al. (2002). Constituents of *Lepidium meyenii* 'maca.' *Phytochemistry*, *59*(1), 105–110.
12. Tellez, M. R., Khan, I. A., Kobaisy, M., Schrader, K. K., Dayan, F. E., & Osbrink, W. (2002). Composition of the essential oil of *Lepidium meyenii* (Walp). *Phytochemistry*, *61*(2), 149–155.
13. Valerio, L. G. Jr., & Gonzales, G. F. (2005). Toxicological aspects of the South American herbs cat's claw (*Uncariato mentosa*) and Maca (*Lepidium meyenii*): A critical synopsis. *Toxicol Rev*, *24*(1), 11–35.
14. McCollom, M. M., Villinski, J. R., McPhail, K. L., Craker, L. E., & Gafner, S. (2005). Analysis of macamides in samples of Maca (*Lepidium meyenii*) by HPLC-UV-MS/MS. *Phytochem Anal*, *16*(6), 463–469.
15. Zhao, J., Muhammad, I., Dunbar, D. C., Mustafa, J., & Khan, I. A. (2005). New alkamides from maca (*Lepidium meyenii*). *J Agric Food Chem*, 53(3), 690–693.
16. Muhammad, I., Zhao, J., Khan, I. A., Maca (Lepidium meyenii)., (2005). *Encyclopedia of Dietary Supplement*, 2nd ed.; Coates, P., Blackman, MR, Cragg, G., Levine, M., Moss, J., White, J., Eds. 522–31.
17. Simona T., Natascia, C., Carotenuto, D., Vassetti, A., Staropoli, A., Mastellone, V., Peretti, V., Ciotola, F., Albarella, S. C., Prete, C. D., Palumbo, V., Esposito, L., Vinale, F., & Ciani, F. (2019). Chemical Analysis of *Lepidium meyenii* (Maca) and Its Effects on Redox Status and on Reproductive Biology in Stallions Molecules, *24*, 1981.
18. Zhou, Y., Li, P., Brantner, A., Wang, H., Shu, X., Yang, J., Si, N., Han, L., Zhao, H., & Bian, B. (2017). Chemical profiling analysis of Maca using UHPLC-ESI-Orbitrap MS coupled with UHPLC-ESI-QqQ MS and the neuroprotective study on its active ingredients. *Sci. Rep*, *7*, 44660.
19. Bennet, R. N., Mellon, F. A., & Kroon, P. A. (2004). Screening crucifer seeds as sources of specific intact glucosinolates using ion-pair High-Performance Liquid Chromatography Negative Ion Electrospray Mass Spectrometry. *J. Agric. Food Chem*, *52*, 428–438.
20. Feng, J., Dou, J., Wu, Z., Yin, D., & Wu, W. (2019). Controlled release of biological control agents for preventing aflatoxin contamination from starch-alginate beads. *Molecules*, *24*, 1858.
21. Fahey, J. W., Zalcmann, A. T., & Talalay, P. (2001). The chemical diversity and distribution of glucosinolates and isothiocyanates among plants. *Phytochemistry*, *56*, 5–51.
22. Bhasin, S., Kretser, D. M., & Baker, H. W. (1994). Clinical review 64: Pathophysiology and natural history of male infertility. *J Clin Endocrinol Metab*, *79*(6), 1525–1529.
23. Patwardhan, B., Warude, D., Pushpangadan, P., & Bhatt, N. (2005). Ayurveda and traditional Chinese medicine: a comparative overview. *Evid. Based Complement Alternat. Med.*, *2*(4), 465–473.
24. , Crimmel, A. S., Conner, C. S., & Monga, M. (2001). Withered Yang: A review of traditional Chinese medical treatment of male infertility and erectile dysfunction. *J. Androl.*, *22*(2), 173–182.
25. Dama, M. S., Akhand, P. S., & Rajender, S. (2010). Nature versus nurture--plant resources in the management of male infertility. *Front Biosci (Elite Ed)*, *2*, 1001–1014.
26. Kapp, B., & Cain, M. (2001). "The neural basis of arousal." In: Smelser, N. & Baltes, P. (Eds.), *The International Encyclopedia of Social and Behavioral Sciences*. Oxford, UK: Elsevier Science.

27. Melisand, M. R., & Argiolas, A. (2013). "Neuropeptides and central control of sexual behavior from the past to the present: A review." *Progress in Neurobiology*, *108*, 80–107.
28. Pfaus, J. G., Damsma, G., Nomikos, G. G., et al. (1990). "Sexual behavior enhances central dopamine transmission in the male rat." *Brain Research*, *530*, (2), 345–348.
29. Hull, E. M., Eaton, R. C., Moses, J., & Lorrain, D. (1993). "Copulation increases dopamine activity in the medial preoptic area of male rats." *Life Sciences*, *52*, (11), 935–940.
30. Hull, E. M., Du, J., Lorrain, D. S., & Matuszewich, L. (1995). "Extracellular dopamine in the medial preoptic area: Implications for sexual motivation and hormonal control of copulation." *Journal of Neuroscience*, *15*(11), 7465–7471.
31. Sato, Y., Wada, H., Horita, H., et al. (1995). "Dopamine release in the medial preoptic area during male copulatory behavior in rats." *Brain Research*, *692*(1–2), 66–70.
32. Giuliano, F., & Rampin, O. (2000). "Central neural regulation of penile erection." *Neuroscience and Biobehavioral Reviews*, *24*(5), 517–533.
33. Waddell, T. G., & Ibach, D. M. (1989). "Modern chemical aphrodisiacs." *Indian Journal of Pharmaceutical Sciences*, *51*(3), 79–82.
34. Moncada, S., Palmer, R. M. J., & Higgs, E. A. (1991). "Nitric oxide: Physiology, pathophysiology, and pharmacology." *Pharmacological Reviews*, *43*(2), 109–142.
35. Snyder, S. H. (1992). "Nitric oxide: First in a new class of neurotransmitters?." *Science*, *257*(5069), 494–496.
36. Forstermann, U., Gorsky, L. D., Pollock, J. S., Schmidt, H. H. H. W., Heller, M., & Murad, F. (1990). "Regional distribution of EDRF/NO-synthesizing enzyme(s) in rat brain." *Biochemical and Biophysical Research Communications*, *168*(2), 727–732.
37. Bredt, D. S., Hwang, P. M., Glatt, C. E., Lowenstein, C., Reed, R. R., & Snyder, S. H. (1991). "Cloned and expressed nitric oxide synthase structurally resembles cytochrome P-450 reductase." *Nature*, *351*(6329), 714–718.
38. Anderson, K. E. (1993a). "Pharmacology of lower urinary tract smooth muscles and penile erectile tissues." *Pharmacological Reviews*, *45*(3), 253–308.
39. Kim, N., Azadzoi, K. M., Goldstein, I.., & Saenz de Tejada, I. (1991). "Anitric oxide-like factor mediates nonadrenergic-noncholinergic neurogenic relaxation of penile corpus cavernosum smooth muscle." *Journal of Clinical Investigation*, *88*(1), 112–118.
40. Holmquist, F., Hedlund, H., & Andersson, K. E. (1991). "L-NG-nitroarginine inhibits non-adrenergic, non-cholinergic relaxation of human isolated corpus cavernosum." *Acta Physiologica Scandinavica*, *141*(3), 441–442.
41. Nehra, A., Colreavy, F., Khandheria, B. K., & Chandrasekaran, K. (2001). "Sildenafil citrate, a selective phosphodiesterase type 5 inhibitor: urologic and cardiovascular implications." *World Journal of Urology*, *19*(1), 40–45.
42. Achike, F. I., & Kwan, C. Y. (2003). "Nitric oxide, human diseases and the herbal products that affect the nitric oxide signaling pathway." *Clinical and Experimental Pharmacology and Physiology*, *30*(9), 605–615.
43. Dohle, G. R., Smit, M., & Weber, R. F. A. (2003). "Androgens and male fertility." *World Journal of Urology*, *21*(5), 341–345.
44. Snyder, P. J. (1984). "Clinical use of androgens." *Annual Review of Medicine*, *35*, 207–217.
45. Hull, E. M., Du, J., Lorrain, D. S., & Matuszewich, L. (1997). "Testosterone, preoptic dopamine, and copulation in male rats." *Brain Research Bulletin*, *44*(4), 327–333.
46. Zvara, P., Sioufi, R., Schipper, H. M., Begin, L. R., & Brock, G. B. (1995). "Nitric oxide-mediated erectile activity is a testosterone dependent event: a rat erection model." *International Journal of Impotence Research*, *7*(4), 209–219.

47. Gauthaman, K., & Adaikan, P. G. (2005). "Effect of *Tribuluster restris* on nicotinamide adenine dinucleotide phosphate-diaphorase activity and androgen receptors in rat brain." *Journal of Ethnopharmacology*, *96*(1–2), 127–132.
48. Du, J., & Hull, E. M. (1999). "Effects of testosterone on neuronal nitric oxide synthase and tyrosine hydroxylase." *Brain Research*, *836*(1–2), 90–98.
49. Liu, Y. C., & Sachs, B. D. (1999). "Erectile function in male rats after lesions in the lateral para gigan to cellular nucleus." *Neuroscience Letters*, *262*(3), 203–206.
50. Tocharus, C., Jeenapongsa, R., Teakthong, T., & Smitasiri, Y. (2005). "Effects of long-term treatment of *Buteasuperba* on sperm motility and concentration." *Naresuan University Journal*, *13*, 11–17.
51. Manosroi, A., Sanphet, K., Saowakon, S., Aritajat, S., & Manosroi, J. (2006). "Effects of *Buteasuperba* on reproductive systems of rats." *Fitoterapia*, *77*(6), 435–438.
52. Cherdshewasart, W., Bhuntaku, P., Panriansaen, R., Dahlan, W., & Malaivijitnond, S. (2008). "Androgen disruption and toxicity tests of *Buteasuperba* Roxb., a traditional herb used for the treatment of erectile dysfunction, in male rats." *Maturitas*, *60*(2), 131–137.
53. Cherdshewasart, W., & Nimsakul, N. (2003). "Clinical trial of *Butensuperba*, an alternative herbal treatment for erectile dysfunction." *Asian Journal of andrology*, *5*(3), 243–246.
54. Tocharus, C., Smitasiri, Y., & Jeenapongsa, R. (2006). "*Buteasuperba* Roxb. enhances penile erection in rats." *Phytotherapy Research*, *20*(6), 484–489.

CHAPTER 4

Matter, Plasma, Quasi Two-Dimensional Au Film, Drug Products Quality, and Circular Economy

FRANCISCO TORRENS[1] and GLORIA CASTELLANO[2]

[1]*Institut Universitari de Ciència Molecular, Universitat de València, Edifici d'Instituts de Paterna, P.O. Box 22085, E-46071 València, Spain, E-mail: torrens@uv.es*

[2]*Departamento de Ciencias Experimentales y Matemáticas, Facultad de Veterinaria y Ciencias Experimentales, Universidad Católica de Valencia San Vicente Mártir, Guillem de Castro-94, E-46001 València, Spain*

ABSTRACT

The use of continuous large-area monolayer molybdenum disulfide was proposed as an adhesion layer, for the growth of ultra-thin metal films by deposition. Via extensive optical, electrical, and structural characterization, ultra-thin gold films on molybdenum disulfide show continuous smooth morphology and Drude plasmonic response at a thickness in 3–4 nm. The results advance using high-quality ultra-thin gold films for manufacturing van der Waals heterostructures, and as transparent electrodes for optoelectronic devices. Chemistry establishes a bridge between the familiar and the fundamental. This account provides an insight into such a bridge but, with any luck, one can see across the fog and understand how a chemist thinks. A growing body of evidence exists that suggests that certain foods, especially sugars, may be addictive, and play an important role in the global obesity epidemic and related diseases, for which reason, more research on the products is encouraged because ignorance can be dangerous and promote only industrial interests. From the industry's viewpoint, it seeks to help processed-food consumption, presenting it as healthy with promotional campaigns with millionaire investments, in order to defend their interests and conquer the least-informed people. The histogram causes a lack of *empathy*, which produces a lack of *action*.

4.1 INTRODUCTION

Setting the scene: some reflections on the idea of matter in physics education, plasma in the stars, the Sun, ultrathin and ultrasmooth gold films on monolayer MoS2, perspectives on the quality attributes of drug products containing nanomaterials, circular bioeconomy (CBE), agro-food system, peroxyacetyl nitrate (PAN) C2H3O5N, nicotine, other tobacco components, sugar, a sweet enemy with bitter reality, unhealthy metabolic behavior, some metaphors that have made history in philosophy and some other metaphors that have made history in science. The following hypothesis (H) appeared on gender in science: Science, not silence.

In earlier publications, it was informed the effects of type, size and elliptical deformation on molecular polarizabilities of model single-wall carbon nanotubes (SWNTs) from atomic increments [1–4], SWNTs periodic properties and table based on the chiral vector [5, 6], calculations on SWNTs solvents, co-solvents, *cyclo*-pyranoses [7–10] and organic-solvent dispersions [11, 12], packing effect on cluster nature of SWNTs solvation features [13], information entropy analysis [14], cluster origin of SWNTs transfer phenomena [15], asymptotic analysis of coagulation–fragmentation equations of SWNT clusters [16], properties of fullerite and symmetric C-forms, similarity laws [17], fullerite crystal thermodynamic characteristics, law of corresponding states [18], cluster nature of nanohorns (SWNHs) solvent features [19], SWNTs (co-)solvent selection, *best* solvents, acids, superacids, host–guest inclusion complexes [20], C/BC_2N/BN fullerenes/SWNTs/nanocones (SWNCs)/SWNHs/buds (SWNBs)/graphemes (GRs) cluster solvation models in organic solvents [21–27], elementary polarizability of Sc/*fullerene*/GR aggregates, di/GR–cation interactions [28], the world of materials from layered ones to conductive two-dimensional (2D) metal–organic frameworks (MOFs) [29–32], porous materials, MOFs, conduction [33], magnetism, polyoxometalates, layered materials, GR [34], mesoporous, GR composite, Li battery, topology, periodicity [35], MOFs, science, pseudo-science and chronobiology [36]. The purpose of the present report is to review some reflections on the idea of *matter* in physics education, plasma in the stars, the Sun, ultrathin and ultrasmooth gold films on monolayer MoS_2, perspectives on quality attributes of drug products containing nanomaterials, CBE, agro-food system, PAN, nicotine, other tobacco components, sugar, a sweet enemy with bitter reality, unhealthy metabolic behavior, some metaphors that made history in philosophy and some other metaphors that made history in science. The aim of this work is to initiate a debate by suggesting a number of questions (Q), which can arise when addressing subjects of the agro-food sector and CBE. It was provided, when possible, answers (A) and H on these ideas, materials, and procedures.

4.2 REFLECTIONS ON IDEA *MATTER* IN PHYSICS EDUCATION: PLASMA IN STARS (SUN)

Knowing and questioning spontaneous teachers' thinking about the *matter* can help overcome simple, disjointed, and unproblematic visions in the construction of scientific knowledge [37]. Examples presented in textbooks, corresponding to distorted visions, result in reductionist images, which are transmitted tacitly in the teaching of matter.

1. Reduction of the general idea of matter existing in Universe to that of ordinary matter present on Earth; *for example*, the physical states of matter are reduced to the four most common ones on Earth.
2. Presentation of the mass and volume of solid and liquid bodies as the only general properties of matter.
3. Abuse of a static representation of the structure of matter, forgetting the consubstantial dynamism of matter itself.
4. Compartmentalized and disjointed presentation of the different levels of organization of matter, which are studied in secondary education (i.e., absence of existing relationships between the different subjects, even in the same, between the macro and microscopic).
5. The lack of interdisciplinary relationships, caused by excessively analytical visions or representations of scientific subjects, makes it difficult to acquire a global and unitary image of the behavior of matter, forgetting that the search for unifying syntheses in the explanation of the world is the true object of science.

Limiting the states of matter to the four most evident on Earth would be as reducing the ordinary matter only to solids and liquids, without considering that gases and vapors present mass and volume. How to exclude from the term *matter* the states that science showed to abound much more in Universe than not in the ordinary matter? It cannot be excluded from the plasma state [e.g., stars (Sun, *cf.* Figure 4.1), magnetic nuclei of planets (Earth), a flame of a candle]. It would be like forgetting the fire element in the Aristotelian paradigm. The idea of the materiality of fire (or plasma) has more than two centuries. The young Kant (1755) opposed a chair of philosophy at the University of Königsberg (East Prussia, now Kaliningrad, Russia) presenting the book *Brief Outline of Certain Meditations on Fire*, where he defined *fire* as: *The matter of fire is but the same ether (the matter of light) compressed in the interstices of the bodies by virtue of the intense force of attraction or adhesion* [38]; That is, it was accepted the existence of great attractive forces between the particles of the fire and ordinary matter, and that fire, light or ether were material. Materiality that people currently accept while saving the ontological and conceptual distances that separate them from that century. They include in a fifth plasma state Sun and the magma of Earth's interior. The solar surface is a kind of *letter soup* with small particles (electrons, protons, neutrons, helions, etc.) whose mass does not exceed 4 a.m.u. because of the strong nuclear interactions (fissions) that raise the temperature to millions of degrees. However, emissions of large quantities of light exist as electromagnetic fields (EMFs) that are transmitted *via* intersideral spaces. It is possible to speak of the sixth state of matter, EMFs, which invades all, which materiality cannot be doubted even though it is extremely subtle and blurred matter. Not surprisingly, it was proven that Sun loses mass at the rate of thousands of tons per second. One can talk about other states of matter that are little known but whose existence astrophysicists proposed [e.g., *dark matter* (DM), *black holes*], which account for 90% of the mass of Universe that is missing to characterize. Recent hypotheses affect the possibility that H_2 and H, in a state of high density, form DM, observed thanks to high gravity effects.

4.3 RESEARCHERS SYNTHESIZE QUASI-TWO-DIMENSIONAL AU FILM

Sub-10 nm continuous metal films are promising candidates for flexible and transparent nanophotonics and optical-electronic (*optoelectronic*) applications. Volkov group showed that monolayer (ML) MoS_2 is a perspective adhesion layer, for the deposition of continuous conductive Au films with a thickness in 3–4nm [39]. Optical properties of continuous Au ultra-thin films (UTFs), deposited on 2D MoS_2 grown by chemical vapor deposition (CVD), are investigated by spectroscopic ellipsometry over a wavelength in 300–3300 nm. Optical losses in UTFs rise with decaying thickness because of the fine-grained structure and the presence of a small number of voids; however, they exhibit metallic properties down to a thickness of 3–4 nm. The atomic-scale MoS_2 interfaces can be transferred to any substrate, and open up opportunities for the creation of metasurfaces and a type of van der Waals (VdW) heterostructures with atomically thin metal layers.

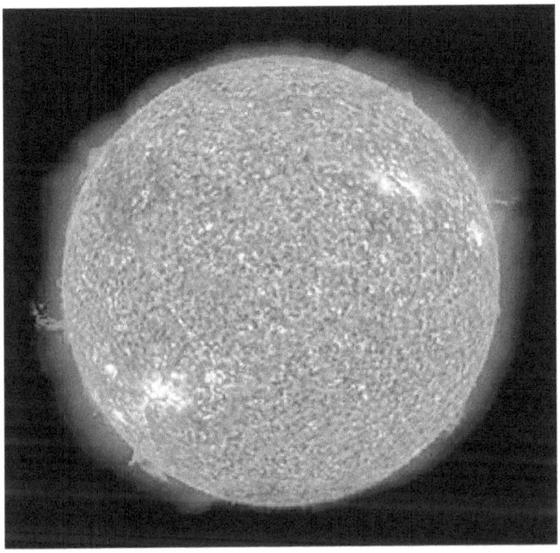

FIGURE 4.1 It cannot be excluded plasma of which existing ones are part of at high temperatures in stars.

4.4 QUALITY ATTRIBUTES OF DRUG PRODUCTS CONTAINING NANOMATERIALS

Zaidi group summarized the contents of a stimulus article produced by the joint subcommittee of *USP* Expert Committees on Dosage Forms, Physical Analysis and Excipients, and relevant comments from a workshop session [40]. The current interest in nanotechnology and its application to the medical field presents historical precedent.

4.5 CIRCULAR BIOECONOMY AND AGRO-FOOD SYSTEM

Guillem Ruiz and Franco Vila organized Day *Circular Bioeconomy and Agro-food System* [41]. Franco Vila proposed H on a *circular economy* (CE).

H1. CE *vs*. linear economy (LE).
H2. Change from LE to CE systems.

- Folkeson proposed questions/answers/hypotheses on the agro-food sector and CBE [42].

Q1. What is CBE?
A1. CBE is a system to maintain the economic value of material products/resources the most time possible and in doing so use resources at maximum and reduce the residues amount to a minimum.
H3. Objective: lesser dependence on non-renewable resources while ensuring food safety.
H4. BE forms part of CE.
H5. BE is not sustainable by definition: Apply to its EC principles.
Q2. How to participate in BE?
H6. Challenges to participate in BE: BE concept is new and not much clear; policies coherence/consistency; market demand is low because products are more expensive than traditional ones; high market demand and supporting policies exist.
Q3. What is BE?
Q4. How can I participate?
Q5. What does it mean in practice?
A5. BE must be accompanied by other strategies.
Q6. How am I going to resolve it?
Q7. What are your potentialities?
Q8. What is your biomass?
Q9. How are you going to develop BE?

4.6 PEROXYACETYL NITRATE $C_2H_3O_5N$

Peroxyacetyl nitrate (PAN) is a residue coming from the exhaust of vehicles and is present in photochemical pollution [43]. The hydrocarbon (HC) fuels that are not completely burned are released to the atmosphere as partially oxidized fragments that include radical $CH_3CO^{\bullet}$ (*cf.* attached fragment at the left of Figure 4.2). $CH_3CO^{\bullet}$ attacks O_2 molecules that surround it and forms $CH_3CO{-}OO^{\bullet}$, a modification of H_2O_2 molecule. During morning pollution, when a lot of NO and sunlight exist, $CH_3CO{-}OO^{\bullet}$ oxidizes NO to NO_2. During the evening, when NO concentration decayed and NO_2 is more abundant, $CH_3CO{-}OO^{\bullet}$ binds NO_2 molecules to form PAN, which is a powerful tear-producing responsible for most eye irritations caused by pollution. It causes tears because the tear ducts reply secreting a saline liquid (that carries antibodies when the invasion is bacterial) when they are invaded in order to wash and get rid of the invader. It is responsible for the damage that pollution causes to vegetation; its high O content causes the oxidation, that is, partial combustion, of any organic matter, that it touches. It damages plants because it causes the tissue at the bottom of the leaves to collapse.

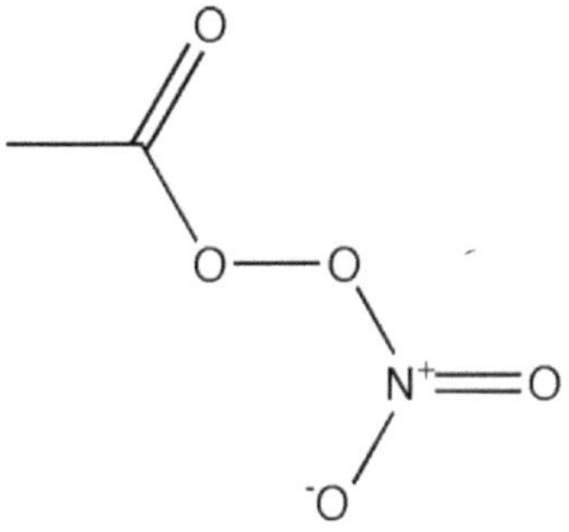

FIGURE 4.2 Molecular structure of peroxyacetyl nitrate.

4.7 IF IT WERE ONLY NICOTINE

A small cylinder with tobacco (cigarette) contains ingredients that originate 7,000 different products, 69 of which are carcinogens, which are divided into: nicotine, tar, and CO [44]. Nicotine (*cf.* Figure 4.3) is a molecule similar to acetylcholine, one of the main natural neurotransmitters, and is greatly responsible for tobacco addiction. Since it arrives in the lungs, it takes only 7s to reach the brain and stimulates the reward areas. It causes insulin resistance (promotes diabetes) and increases the aggregation of platelets (as opposed to aspirin). Tobacco is a relatively important source of $CO_{(g)}$, the origin of which comes from the incomplete combustion of a large number of substances and that is, for example, responsible for the asphyxiation deaths in houses with brash or poorly ventilated garages. It has the property of joining hemoglobin with a force 200′ greater than O_2. The tar is made up of all the particles that remain in certain filters after extracting nicotine and water: It is the obscure and viscous substance deposited in the lungs and contains a large part of the carcinogenic components, for example, polycyclic aromatic hydrocarbons (HCs, present also in petroleum), nitrosamines and benzopyrenes. A lot of fine particles (measuring less than 2.5 mm) exist like those produced in car engines, which penetrate deep into the lungs, causing a number of side effects [e.g., respiratory, cardiovascular diseases (CVDs)]. Tobacco contains an innumerable list of substances that are harmful to people's health, for example, toluene, an HC that exists naturally in petroleum. It is used as an additive and solvent in gasoline. In the body, it acts as an irritant and is toxic to the central nervous system (CNS). Formaldehyde and acetaldehyde exist, present in natural products, and used for the manufacture of plastics and paints. Formaldehyde is considered a Group-1 (the strongest) carcinogen. Acetaldehyde occurs naturally during the combustion of tobacco but increases with the addition of sugars, and increases addiction to nicotine. Another component is acrylamide, toxic for CNS. Tobacco contains As and Cd. As was associated with a greater risk of diabetes, CVDs, and neurotoxicity, and is a Group-1 carcinogen. Cd is used in the manufacture of batteries and was linked to lung, CVDs, and kidney diseases. It is a Group-1 carcinogen. ^{210}Po is a radioactive element in tobacco. It is a carcinogen and, at high doses, causes death by poisoning. Notice the use of additives, for example, NH_3, to improve the consistency and taste of tobacco, but also to reduce acidity and make it possible to absorb more nicotine to enhance the addictive effect. Companies introduced vitamin-A compounds in an unsuccessful attempt to diminish the cancerous power of tobacco.

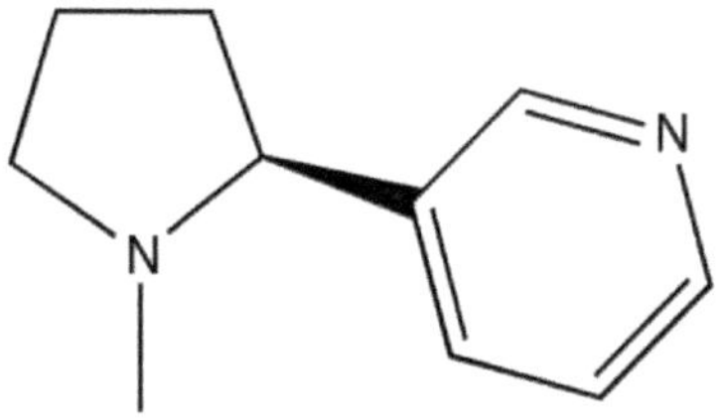

FIGURE 4.3 Molecular structure of nicotine.

4.8 SUGAR: SWEET ENEMY, BITTER REALITY, UNHEALTHY METABOLIC BEHAVIOR

Some food panics appeared *via* history (*cf.* Table 4.1) [45].

The refined sugar (any monosaccharide) does not require steps of digestive metabolic treatment, but they are absorbed direct and massively after being ingested, which causes important glucose peaks in the blood (*cf.* Figure 4.4). A growing body of evidence exists that suggests that certain foods, particularly sugars, may be addictive and play an important role in the global obesity epidemic and related diseases. Carbohydrates (CH_2O) show unhealthy metabolic behavior, as discussed below.

TABLE 4.1 Some Food Panics

Food panic
Tobacco
Alcohol
Saturated fats
Sugars (flavorings)
Vegetable *trans*-fats
Red meat
Panga
Palm oil

1. A different way of absorbing sugars according to their origin.
2. The strong glycemic peaks facilitate obesity because they favor the rise in fat reserves.
3. Hyperglycemic peaks are followed by hypoglycemic tendencies that induce eating again.
4. Glucose peaks appear to be like addictive generators.
5. Consumption of vitamins.

FIGURE 4.4 Sugar: Sweet enemy with bitter reality. Equation: Sugar = Poison. Source: Ref. [45].

4.9 SOME METAPHORS THAT HAVE MADE HISTORY IN PHILOSOPHY

Strauss published an introduction to philosophy [46]. Table 4.2 gives some metaphors that have made history in philosophy [47].

TABLE 4.2 Some Metaphors that have Made History in Philosophy

Metaphor
(Plato, *Teeteto*). Human memory: A block of wax.
(Plato, *Teeteto*). Human mind: An aviary gradually occupied by different classes of birds, which correspond to the varieties of knowledge.
Humans are born to die.
We all die in the end.
(Cervantes). As much true, Sancho, as someday we will be dead.
(J. M. Keynes). In the long run, we are all dead!
(Jorge Wagensberg). Death is the most surprising of all the predictable notices.
(Jorge Wagensberg). One has just died completely when the last one who has known him alive dies.
(Marlowe, *Doctor Faustus*, 1590). Faustus' pact (the use that people do of fossil fuels).
(Fidel Castro). Biofuel will increase the price of food
(Darwin, 1837). Tree of life.
(Louis L'Amour). The more one learns the more he understands his ignorance.
(Albert Einstein). The more I learn, the more I realize how much I don't know.
(Javier Sábada). The delimitation that supposes the beach between the ocean of the unattainable and the rationality of the land.
Value: A monetary metaphor used for the intellectual or philosophical value.
Economic forces: The ones that are governed by the benefit
(Mao Zedong). The best poems are written in a blank book.
(Jorge Wagensberg). Science is a fiction of reality.
(Jorge Wagensberg). To educate is to favor addiction to intellectual joy.
(Jorge Wagensberg). Learning has three phases: stimulus, conversation, and understanding, and with each, there is the opportunity for intellectual enjoyment.
(Jorge Wagensberg). A good teacher does not hide the paradoxes but he looks for them.
(Jorge Wagensberg). Knowledge without criticism is more worrisome than criticism without knowledge.
(Jorge Wagensberg). The greatness of science is that it is capable of understanding without needing to intuit.

TABLE 4.2 *(Continued)*

(Jorge Wagensberg). The greatness of art is that it is capable of intuiting without needing to understand.
(Jorge Wagensberg). If it were not for the crises, we would still be all bacteria.
(Fernando Senent). If the Earth were round, we would be fishes.

4.10 SOME METAPHORS THAT HAVE MADE HISTORY IN SCIENCE

A tool, clustered, regularly interspaced, short palindromic repeat (CRISPR), allows scientists to modify the genome like *editors* that *rewrite the book of life*, which is one of the literary figures about the advances of genetics [48]. Since the Human Genome Project (HGP, 1990), metaphors followed their vital cycle: they are born, settle into, die, and are replaced. Whereas in the 2000s, DNA was a *map to decipher*, now it is a *book*. The images mark the evolution of science. Keller noticed that the ways in which people talk about science objects are not simply determined by empirical evidence, but they actively influence the type of proof that people search and how a research is driven [49]. Table 4.3 lists some metaphors that have made history in science. Lakoff and Johnson analyzed how the metaphors that people use to model their perception, thought, and actions [50]. Sontag noticed that the utilization of war language, in cancer description, can present an unbearable weight in which suffer from it [51]. Lizcano noticed that the problem appears when, because of their repeated application, the metaphors turn invisible, natural [52]. The periodic table of the elements (PTE) is related to the literature [53]. Salvador Liern informed on analogical knowledge and metaphors in biotechnology discourse [54].

TABLE 4.3 Some Metaphors That Have Made History in Science

Concept	Metaphor
Electricity	Current, electrons *flow* like water on a riverbed
Telephone system	Neural network
Internet	Information superhighway, Cloud
Atom	Solar system in miniature, satellites
Elementary particles	Bricks of matter
Spacetime	Weave of cosmos
Higgs' boson	Particle of God
Plasma	Letter soup of electrons, protons, neutrons, helions, etc.

TABLE 4.3 *(Continued)*

Concept	Metaphor
Continuous light source	Rainbow
Discontinuous light source	Barcode
Chemical element	Letter of alphabet
Periodic table of the elements	Genealogical tree
Sustainable chemistry	Green chemistry
Agricultural biotechnology	Green biotechnology
Medicinal biotechnology	Red biotechnology
Industrial Biotechnology	White biotechnology
Marine biotechnology	Blue biotechnology
Nature	Book was written in mathematical language
Men play God	Men *play God*
Brain	Human-computer
Mind	Brain software
Human mind	An aviary gradually occupied by different classes of birds, which correspond to the varieties of knowledge (Plato)[a]
Human memory	A block of wax (Plato)[a]
Amino acids	Bricks of life
Genome	Code, map, manual, identity card, genetic passport
Mosaicism	Existence of cells with different genetic and chromosomal content within the same organism, mosaic-like
Transgenic farming	*Transgenic* farming
Cloning	An attempt to *patent* a living being/produce babies with a warranty to be free of health-related problems with the result being *monstrous*/*miraculous*
Genetic promiscuity	Genetic *promiscuity*
Genome edition technique	Text processor
CRISPR	Genetic *cut–paste*, molecular scissors, molecular scalpel, cut out, paste and color
Mitochondrion	Cell power plant
Ribosome	Assembly line, Turing machine
Immune system	Defense army, battlefield, arms race
Baculoviruses hijack caterpillar cells	Baculoviruses *hijack* caterpillar cells
Bacteria invented sex	Bacteria *invented* sex
Drug	Magic bullet

TABLE 4.3 *(Continued)*

Concept	Metaphor
Iatrogenesis	*Iatrós*=medical origin
Antibiotic	Weapon against bacteria
Radiologists that avoid destroying healthy tissue	Soldiers conquering a territory without running into the mines there hidden (Gick and Holyoak, 1980)[b]
Virus	Invading enemy
Enzyme and substrate	Lock and key
Cri du chat or cat-cry syndrome	Disease in which the child patient cries and screams in a high-pitched voice
Protein, nucleic acid, etc.	Biological target
Gatekeeper protein	*Gatekeeper* protein
Molecular sieve	Molecular *sieve*
Nanopores *screen* nucleotides	Like *fruit sorters*
DNA library	DNA *library*
Body	Cell society, machine
Heart	Blood pump
Epydermolysis bullosa	Butterfly skin
Nervous system	Electric net
Feelings	Cupid with his quiver of laced arrows, the Furies of passion, the demon of jealousy
Cell	*Cell* of a honeycomb
A cell develops depressive behavior	A cell develops *depressive behavior*
Stem cell	*Root cell*
Mitosis	*Mitos*=thread
Apoptosis	*Apoptosis*=fall, cell suicide; a natural killer cell *convinces* other cells to commit suicide
Earth	Living organism; the Mother–Earth goddess Gaia, who naturally regenerates
Animal	Medieval bestiaries, classic cautionary tales
The chrysalides of cabbage butterflies are protein factories	The chrysalides of cabbage butterflies are *protein factories*
Plant	Biofactory
Plant stress	Plant *stress*
Weed	*War* on weeds: knapweed
Algae are persuaded to generate electricity	Algae are *persuaded* to generate electricity

TABLE 4.3 *(Continued)*

Concept	Metaphor
Evolution	Society based on competence, tree diagram (Darwin, 1837), marine coral (Darwin, 1837)
Symbiosis	Society based on collaboration
Meme	Gene
Achromatic spindle	*Achromatic spindle*
Circadian rhythm	*Circa*=around, *dies*=day
Biological clock	Biological *clock*
Internal clock	Internal *clock*
Jet lags slow down/speed up the ticking of our biological clock to reprogram our daily biorhythm	Jet lags *slow down/speed up* the ticking of our *biological clock* to *reprogram* our daily biorhythm
Biomimetics	Consideration of the structure/function of certain biosystems as models that are useful for designing/engineering materials/machines

[a]*Plato, Teeteto.*

[b]*Gick and Holyoak [55].*

4.11 FINAL REMARKS

From the present results and discussion, the following final remarks can be drawn.

1. To use continuous large-area monolayer MoS_2 was proposed as an effective adhesion layer, for the growth of ultra-thin metal films by a standard deposition method. By extensive optical, electrical, and structural characterization, ultra-thin Au films on MoS_2 exhibit continuous smooth morphology and Drude plasmonic response at thickness down to 3–4 nm. The obtained results advance the use of high-quality ultra-thin Au films for fabrication of van der Waals heterostructures, and as transparent electrodes for optoelectronic devices.
2. Chemistry establishes a bridge between the familiar and the fundamental. This report provides a general view of such a bridge but, hopefully, one can see across the fog and understand how a chemist thinks.
3. A growing body of evidence exists that suggests that certain foods, particularly sugars, may be addictive, and play an important role in the global obesity epidemic and related diseases, for which reason,

more research on the products needs to be encouraged, because ignorance can be dangerous and promote only industrial interests, for example, the effects of tobacco. From the point of view of the industry, it seeks to endorse the consumption of processed food, presenting it as healthy with promotional campaigns with millionaire investments, in order to defend their interests and conquer the least-informed people.

4. In the graphical representation of molecular properties, the histogram (basically unchanged for 200 years) causes a lack of *empathy*, which produces a lack of action.

ACKNOWLEDGMENTS

The authors thanks the support from Fundacion Universidad Catolica de Valencia San Vicente Martir (Project No. 2019-217-001UCV).

KEYWORDS

- **cardiovascular diseases**
- **central nervous system**
- **electromagnetic fields**
- **respiratory**

REFERENCES

1. Torrens, F., (2003). Effect of elliptical deformation on molecular polarizabilities of model carbon nanotubes from atomic increments. *J. Nanosci. Nanotech., 3*, 313–318.
2. Torrens, F., (2004). Effect of size and deformation on polarizabilities of carbon nanotubes from atomic increments. *Future Generation Comput. Syst., 20*, 763–772.
3. Torrens, F., (2004). Effect of type, size and deformation on polarizability of carbon nanotubes from atomic increments. *Nanotechnology, 15*, S259–S264.
4. Torrens, F., (2006). Corrigendum: Effect of type, size and deformation on polarizability of carbon nanotubes from atomic increments. *Nanotechnology, 17*, 1541–1541.
5. Torrens, F., (2004). Periodic table of carbon nanotubes based on the chiral vector. *Internet Electron. J. Mol. Des., 3*, 514–527.
6. Torrens, F., (2005). Periodic properties of carbon nanotubes based on the chiral vector. *Internet Electron. J. Mol. Des., 4*, 59–81.

7. Torrens, F., (2005). Calculations on *cyclo*pyranoses as co-solvents of single-wall carbon nanotubes. *Mol. Simul., 31*, 107–114.
8. Torrens, F., (2005). Calculations on solvents and co-solvents of single-wall carbon nanotubes: *Cyclo*pyranoses. *J. Mol. Struct. (THEOCHEM), 757*, 183–191.
9. Torrens, F., (2005). Calculations on solvents and co-solvents of single-wall carbon nanotubes: Cyclopyranoses. *Nanotechnology, 16*, S181–S189.
10. Torrens, F., (2005). Some calculations on single-wall carbon nanotubes. *Probl. Nonlin. Anal. Eng. Syst., 11*(2), 1–16.
11. Torrens, F., (2006). Calculations of organic-solvent dispersions of single-wall carbon nanotubes. *Int. J. Quantum Chem., 106*, 712–718.
12. Torrens, F., & Castellano, G., (2005). Cluster origin of the solubility of single-wall carbon nanotubes. *Comput. Lett., 1*, 331–336.
13. Torrens, F., & Castellano, G., (2007). Cluster nature of the solvation features of single-wall carbon nanotubes. *Curr. Res. Nanotech., 1*, 1–29.
14. Torrens, F., & Castellano, G., (2007). Effect of packing on the cluster nature of C nanotubes: An information entropy analysis. *Microelectron. J., 38*, 1109–1122.
15. Torrens, F., & Castellano, G., (2007). Cluster origin of the transfer phenomena of single-wall carbon nanotubes. *J. Comput. Theor. Nanosci., 4*, 588–603.
16. Torrens, F., & Castellano, G., (2007). Asymptotic analysis of coagulation-fragmentation equations of carbon nanotube clusters. *Nanoscale Res. Lett., 2*, 337–349.
17. Torrens, F., & Castellano, G., (2008). Properties of fullerite and other symmetric carbon forms: Similarity laws. *Symmetry Cult. Sci., 19*, 341–370.
18. Torrens, F., & Castellano, G., (2010). Fullerite crystal thermodynamic characteristics and the law of corresponding states. *J. Nanosci. Nanotechn., 10*, 1208–1222.
19. Torrens, F., & Castellano, G., (2010). Cluster nature of the solvent features of single-wall carbon nanohorns. *Int. J. Quantum Chem., 110*, 563–570.
20. Torrens, F., & Castellano, G., (2011). (Co-)solvent selection for single-wall carbon nanotubes: *Best* solvents, acids, superacids and guest–host inclusion complexes. *Nanoscale, 3*, 2494–2510.
21. Torrens, F., & Castellano, G., (2012). *Bundlet* model for single-wall carbon nanotubes, nanocones and nanohorns. *Int. J. Chemoinf. Chem. Eng., 2*(1), 48–98.
22. Torrens, F., & Castellano, G., (2013). Solvent features of cluster single-wall C, BC_2N and BN nanotubes, cones and horns. *Microelectron. Eng., 108*, 127–133.
23. Torrens, F., & Castellano, G., (2013). Corrigendum to: Solvent features of cluster single-wall C, BC_2N and BN nanotubes, cones and horn*s*. *Microelectron. Eng., 112*, 168.
24. Torrens, F., & Castellano, G., (2013). *Bundlet* model of single-wall carbon, BC_2N and BN nanotubes, cones and horns in organic solvents. *J. Nanomater. Mol. Nanotech., 2*, 1000107–1–9.
25. Torrens, F., & Castellano, G., (2013). C-nanostructures cluster models in organic solvents: Fullerenes, tubes, buds and graphenes. *J. Chem. Eng., 7*, 1026–1035.
26. Torrens, F., & Castellano, G., (2014). Cluster solvation models of carbon nanostructures: Extension to fullerenes, tubes and buds. *J. Mol. Model., 20*, 2263–1–9.
27. Torrens, F., & Castellano, G., (2014). Cluster model expanded to C-nanostructures: Fullerenes, tubes, graphenes and their buds. *Austin J. Nanomed. Nanotech., 2*(2), 7–1–7.
28. Torrens, F., & Castellano, G., (2013). Elementary polarizability of Sc/*fullerene*/*graphene* aggregates and di/graphene–cation interactions. *J. Nanomater. Mol. Nanotech., S1*, 001–1–8.

29. Torrens, F., & Castellano, G., (2018). Conductive layered metal–organic frameworks: A chemistry problem. *Int. J. Phys. Study Res., 1*(2), 42.
30. Torrens, F., & Castellano, G., (2019). Conductive two-dimensional nanomaterials: Metal-organic frameworks. *J. Appl. Phys. Nanotech., 2*(2), 52.
31. Torrens, Z. F., (2019). From layered materials to bidimensional metal-organic frameworks. *Nereis, 2019*(11), 65–80.
32. Torrens, F., & Castellano, G. (2020). World of conductive two-dimensional metal–organic frameworks. In: Yaser, A. Z., Khullar, P., & Haghi, A. K., (eds.), *Green Materials and Environmental Chemistry: New Production Technologies, Unique Properties, and Applications*. Apple Academic–CRC: Waretown, NJ.
33. Torrens, F., & Castellano, G. (2020). Porous materials, metal–organic frameworks and conduction. In: Esteso, M. A., Ribeiro, A. C. F., & Haghi, A. K., (eds.), *Chemistry and Chemical Engineering for Sustainable Development: Best Practices and Research Directions*. Apple Academic–CRC: Waretown, NJ.
34. Torrens, F., & Castellano, G. (2020). Magnetism, polyoxometalates, layered materials and graphene. In: Kulkarni, S., Rawat, N. K., & Haghi, A. K., (eds.), *Green Chemistry and Green Engineering: Processing, Technologies, Properties, and Applications.* Apple Academic–CRC: Waretown, NJ.
35. Torrens, F., & Castellano, G. (2020). Mesoporous, graphene composite, Li battery, topology, and periodicity. In: Kulkarni, S., Rawat, N. K., & Haghi, A. K., (eds.), *Green Chemistry and Green Engineering: Processing, Technologies, Properties, and Applications*. Apple Academic–CRC: Waretown, NJ.
36. Torrens, F., & Castellano, G. (2020). MOFs, Science, Pseudoscience, and Chronobiology. In: Kulkarni, S., Rawat, N. K., & Haghi, A. K., (eds.), *Green Chemistry and Green Engineering: Processing, Technologies, Properties, and Applications*. Apple Academic–CRC: Waretown, NJ.
37. Furió, M. C., & Furió, G. C., (2016). Algunes reflexions sobre la idea de matèria. *DAUALDEU, 2016*(11), 11–13.
38. Kant, I., (1992). *Opúsculos de Filosofía Natural*. Alianza Editorial: Madrid, Spain.
39. Yakubovsky, D. I., Stebunov, Y. V., Kirtaev, R. V., Ermolaev, G. A., Mironov, M. S., Novikov, S. M., Arsenin, A. V., & Volkov, V. S., (2019). Ultrathin and ultra smooth gold films on monolayer MoS_2. *Adv. Mater. Interfaces, 6*, 1900196-1-6.
40. Hickey, A., Akala, E., Block, L., Hammond, J., Hussain, M., Kleintop, B., Lewen, N., et al., (2019). Perspectives on quality attributes of drug products containing nanomaterials. *Pharm. Technol. Eur. APIs, Excipients, and Manufacturing 2019*, s12–s14.
41. Guillem, R. J. V., & Franco, V. J. A., (2019). *Book of Abstracts Bioeconomía Circular y Sistema Agroalimentaro; Fundación Valenciana de Estudios Avanzados (FVEA)*. València, FVEA: València, Spain.
42. Folkeson, C., (2019). *Book of Abstracts Bioeconomía Circular y Sistema Agroalimentaro; Fundación Valenciana de Estudios Avanzados (FVEA).* València, Spain, FVEA: València, Spain, 2019; O-1.
43. Atkins, P., (2003). *Atkins' Molecules*. Cambridge University: Cambridge.
44. Bolufer, C., (2016). If only it were nicotine. *DAUALDEU, 2016*(11), 35.
45. Fuentes, D., (2017). *Sugar, sweet enemy with bitter reality. DAUALDEU, 2017*, (12), 5–7.
46. Strauss, L., (2018). *Xenophon's Taste: An Introduction to Philosophy*. Biblioteca Nueva: Madrid, Spain.

47. Pascual, S., (2017). Why Philosophy? Neopàtria: Alzira, València, Spain.
48. Kukso, F., (2016). *CRISPR's new genomic revolution makes metaphors flourish. Anuario SINC, 2016*, 120–123.
49. Keller, E. F., (1996). *Refiguring Life: Metaphors of Twentieth-Century Biology*. Columbia University: New York, NY.
50. Lakoff, G., & Johnson, M., (2003). *Metaphors We Live By*. University of Chicago; London, UK.
51. Sontag, S., (1978). *Illness as Metaphor*. Farrar, Straus and Giroux: New York, NY.
52. Lizcano, E., (2006). *Metaphors That Think About Us: About Science, Democracy and other Powerful Fictions*. Bajo Cero–Traficantes de Sueños: Madrid, Spain.
53. Baldridge, K., & Abendroth, T., (2008). *Mixing with Merry Gold: A Faity Tale*.Wiley–VCH: Weinheim, Germany.
54. Salvador, L. V., (2019). On analogical knowledge: Metaphors in biotechnology discourse. *Mètode Sci. Stud. J., 9*, 47–51.
55. Gick, M. L., & Holyoak, K. J., (1980). Analogical problem solving. *Cog. Psychol., 12*, 306–355.

CHAPTER 5

Deep Insights the Privileged Herbotherapeutic Inventory for Hemorrhoidal Treatment

PARAS KOTHARI,[1] ANKITA SONI,[1] and DEBARSHI KAR MAHAPATRA[2]

[1]*Department of Pharmaceutics, Gurunanak College of Pharmacy and Technical Institute, Nagpur – 440026, Maharashtra, India*

[2]*Assistant Professor, Department of Pharmaceutical Chemistry, Dadasaheb Balpande College of Pharmacy, Nagpur – 440037, Maharashtra, India, E-mail: mahapatradebarshi@gmail.com*

ABSTRACT

Hemorrhoids, also known as piles, are vascular structures in the anal canal which help with stool control. They become a disease when swollen or inflamed. Generally, there are two types of hemorrhoids and the sign and symptoms depend on the type present. Internal hemorrhoids are painless, often result in rectal bleeding when defecating. External hemorrhoids result in pain and swelling in the anal area. Male and female are both equally affected most often between 45 and 65 years of age, the condition lasts for a few days. Several factors that lead to the diseased condition include constipation or diarrhea, lack of exercise, low fiber diets, increased intra-abdominal pressure, genetics, aging. Management of pain, inflammation, bleeding is a goal of treatment in hemorrhoids. The present chapter gives information about the pathophysiology and detailed review of treatment using herbal formulation and reveals their different pharmacological activities such as anti-inflammatory, analgesic, anti-oxidant, anti-microbial, immunosuppressant, and laxative properties.

5.1 INTRODUCTION

Hemorrhoids are nowadays getting a very common disease, mainly due to the lifestyle changes that are happening. It affects millions of people in the world and it is a major socioeconomic problem. There are many reasons for hemorrhoids including constipation and prolonged straining. In hemorrhoids, there is substantial enlargement and distal displacement of normal anal cushions [1]. It is the 4th leading disease in the world. It is reported maximum in the age group of 45 to 65 years of age. It has a low mortality rates, it is treatable but has a high impact on the quality of life of the patient and social problems as well as mental insecurity. It can be managed surgically as well as non-surgically [2].

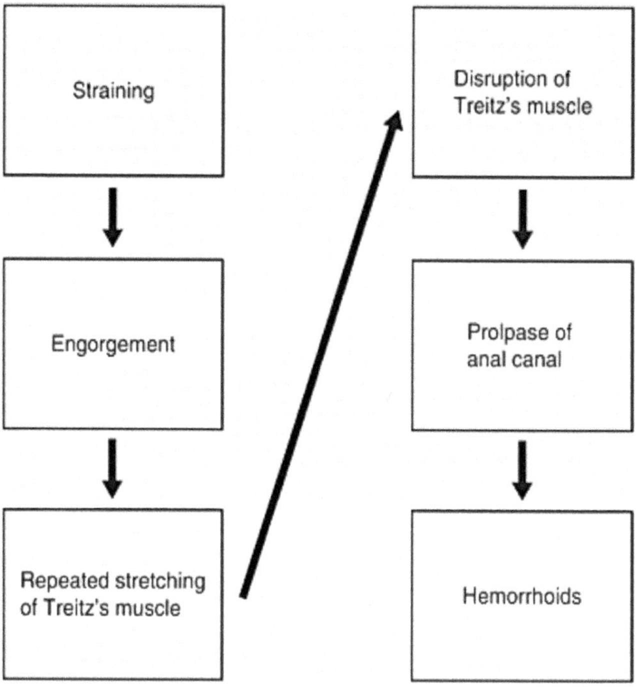

FIGURE 5.1 Stages of hemorrhoids development.

The reason behind the hemorrhoids is still unknown but the studies at Milwaukee VA medical center found that diarrhea is the reason. But many researchers associate constipation with hemorrhoids. They believe that constipation eventually leads to constipation. However, there is no evidence to prove it [3]. The probable mechanism may be that there is an increase in anal sphincter pressure in case of hemorrhoids which further leads to disruption of the anal cushions (Figure 5.1). This continuous pressure cycle leads to increased sphincter pressure and causes distension [4].

Hemorrhoids are located above or below the mucocutaneous junction of the anal canal (Figure 5.2). They bulge into the lumen, sometimes protruding through the anal opening. They are present in every age group from the newborn to the elderly. The important parts of hemorrhoids are the lining, the blood vessels surrounded by a connective tissue stroma, and the anchoring system [5].

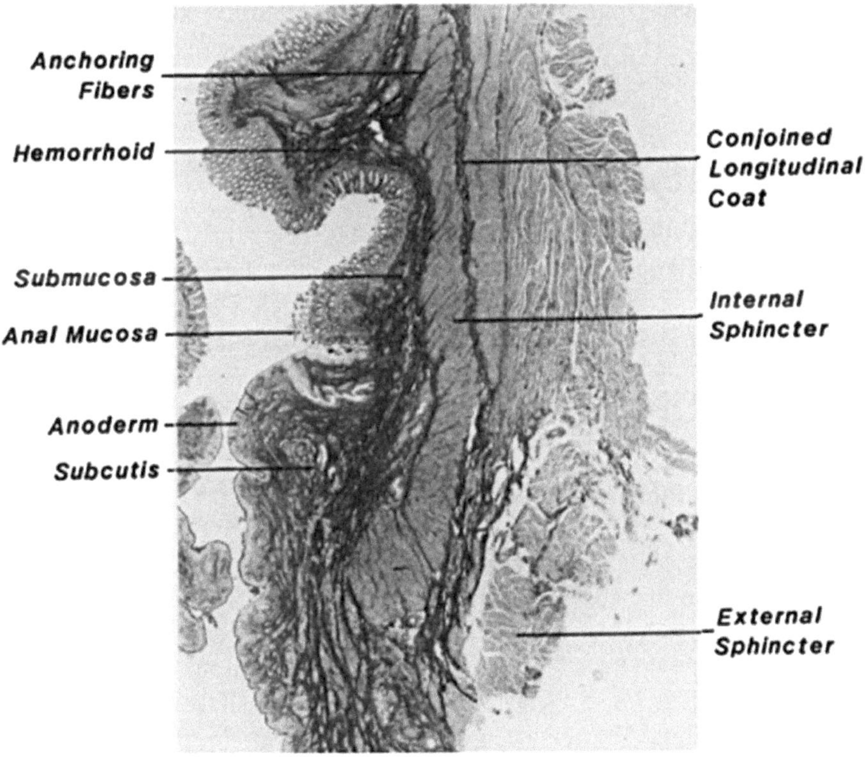

FIGURE 5.2 Anatomical features of hemorrhoids.

There are two types of hemorrhoids:

1. **Internal hemorrhoids:** Internal hemorrhoids are located above the dentate line. They are covered with columnar epithelium and

innervated by visceral nerve fibers that are not associated with pain. Internal hemorrhoids are graded based on the extent of purpose.

2. **External hemorrhoids:** External hemorrhoids are located below the dentate line. They are covered with squamous epithelium and are acutely painful.

5.2 HERBAL TREATMENT FOR HEMORRHOIDS

The least studied subject is hemorrhoids with herbal perspective. There are various treatments available for hemorrhoids but there is less faith in herbal medicine as compared to other medicine. The effect takes place gradually but has substantial improvement in various patients there are various herbal preparations used for the treatment of hemorrhoids some of them are rue care oil, olive oil, white lupine, white dammar, mint, and sesame seed oil [6].

5.3 TREATMENT APPROACHES

The chapter highlights (Figure 5.3) the role of ethnopharmacological traditional knowledge of several plants with potential anti-hemorrhoid activities like *Balsamodendron mukkul*, Shilajit, *Emblica officinalis*, *Berberis aristate*, *Terminalia chebula*, *Cassia Fistula*, *Bauhinia variegate*, *Mesua ferrea*, *Thymus vulgaris*, *Allium ampeloprasum*, *Melia azedarach*, *Myrtle communis*, etc., in Table 5.1.

5.4 OTHER PREVENTIVE MEASURES

The main reason for developing the hemorrhoids into disease conditions is the hardness of the stool. Apart from the medication used, there are certain preventive measures that inhibit the disease progression

a) Fibrous foods – Fibers make the stool soft and allow passing easily, so these should be included in a diet. These foods absorb more water and keep stools soft and regular. Improves bleeding from hemorrhoids.
b) Fluids – Drink 8 to 10 glasses of water and other liquids.
c) Don't stop the urge – One should not avoid going for the toilet, otherwise, the stool will become dry and bleeding may start.
d) Exercise – Regular exercise and maintaining body health will reduce pressure on veins.

e) Avoid long periods of sitting

FIGURE 5.3 Mother Nature's gift to mankind for treating hemorrhoids.

TABLE 5.1 Description, Uses, Biological Sources, and Chemical Constituents of Anti-Hemorrhoidic Plants

Plant	Uses	Biological source	Chemical constituents
Balsamod-endron mukul	1. Cardiac disorders 2. Skin diseases 3. Urinary disorders 4. Hepatic disorders 5. Obesity 6. Gout 7. Leprosy 8. Asthma 9. Bronchitis 10. Hemorrhoids 11. Fever 12. Wounds 13. Anemia 14. Diabetes	*Balsamodendron mukul* hook ex. Stocks, an indigenous plant, belongs to the family Bursereaceae and is synonymous with *Commiphora Mukul.*	1. Ellagic acid 2. Ferric acid 3. Comphorone A and membrane alcohol 4. Allylcembrol 5. Flavones 6. Amino acids 7. Linloeic acid
Shilajit	1. Astringent	Shilajeet is a gelatinous substance that is secreted from the side of the mountains when they have become heated by the rays of the sun in the months of Jyaishta and Ashadha.	Benzoic acid

Plant	Uses	Biological source	Chemical constituents
	2. Anti-inflammatory		Hippuric acid
	3. Anti-ulcerogenic		Fatty acids
	4. Anti-diabetic activity		Resin waxy materials
	5. Memory enhancement		Gums
	6. Anxiolytic activity		Albuminoids
	7. Anti-stress activity		Vegetable matter
	8. Anti-allergic activity		Humus (60–80%)
			Ichthyol
			Ellagic acid
			Resin
			Triterpenes
			Sterol
			Aromatic carboxylic acid
Melia azadirachta	1. Antifungal	*Melia azadirachta* belongs to Meliaceae family of flowering plants order of Saphindales. It comprises 51 genera and about 575 species of trees and rarely shrubs or sometimes shrublets, monopodial or sympodial usually deciduous, less often monoecious or polygamodioecious, native to tropical and subtropical regions. Most members of the family have more leaves, with the leaflets arrange in the form of a feather, and branched flower clusters.	1. Flavonoids

TABLE 5.1 *(Continued)*

Plant	Uses	Biological source	Chemical constituents
	2. Anti-bacterial	The fruit is fleshy and colored or a leathery capsule, *Melia azedarach* L. is a small to medium-size deciduous tree or shrub of 5–15 m in height, it is a close relative of neem.	2. Terpenoids
	3. Hepatoprotective		3. Steroids
	4. Antioxidant		4. Acids
	5. Anti-fertility		5. Anthraquinone
	6. Anthelminthic		6. Kampherol
	7. Antipyretic		7. Quercetin
	8. Cytotoxic		8. Stigma sterol
	9. Vermifuge		9. β-sitosterol
	10. Expectorant		10. β-carotene
	11. Leprosy		11. heptadecane
	12. Anti-parasitic and antifungal		12. 3,5,11,15- tetra methyl -2-hexadecenol
Emblica officinalis	1. Laxative action	*Emblica officinalis*, synonym Amla	1. Tannins
	2. Antioxidant activity	Genus *Embilica* belonging to Family *Euphorbiaceae*),	2. Alkaloids
	3. Anti-tumor activity	Is extensively found all over India, as well as Sri Lanka, Malaya, China, Pakistan, and Bangladesh. The fruits of the plant are fleshy with a sour, astringent taste and are consumed raw	3. Phenolic acids

Plant	Uses	Biological source	Chemical constituents
	4. Anti-inflammatory activity 5. Anti-hypercholesterol 6. Inhibit clastogenicity 7. Inhibit Mutagenicity 8. High vitamin C content 9. Immunomodulator 10. Immunosuppressive	Taxonomy: Kingdom: Plantae Division: Angiospermae Class: Dicotyledonae Order: Gerniales Family: Euphorbiaceae Genus: Embilica Species: Officials Geartn	4. Amino acids 5. Vitamin C 6. Flavonoids 7. Kaemferol 8. Gallic acid 9. Ellagic acid 10. Chebulic acid 11. Quercetin 12. Rhamnopyronoside 13. Corilagin
Berberis aristata	1. Skin diseases 2. Menorrhagia 3. Diarrhea 4. Cholera 5. Jaundice 6. Eye and ear infections 7. Infected wounds 8. Ulcers 9. Cauterization 10. Hepatoxicity	*Berberis aristata* DC. Family Berberidaceae, known locally as Chitra and Dar-E-Hald in Urdu, Dar-Hald in Hindi is a spinous shrub native to mountains parts of north India and Nepal these shrubs are distributed throughout the Himalayas, from Bhutan to Kunawar. The plant is an erect glabrous spinescent shrub, 3–6 m in height with obovate to elliptic, subacute to obtuse, entire or toothed leave (3.8–10 × 1.5–3.3 cm long), flowers are yellow in corymbose racemes, petiole distinct up to 4 mm, inflorescence a single drooping raceme (2.5–7.5 cm long), and dense-flowered.	1. Protoberberine (karachine) 2. Aromoline 3. Oxyberberine 4. Quercetin 5. Rutin 6. Meratin 7. Caffeic acid 8. Chlorogenic acid 9. Pakistanine 10. Taxilamine 11. Tetrahydropalmatine 12. Tetrahydroberberine 13. Palatine

TABLE 5.1 *(Continued)*

Plant	Uses	Biological source	Chemical constituents
			14. Palatine Cl
			15. N-docasane
Terminalia chebula	1. Gentle laxative	It consists of dried ripe fruit, root, bark of plant *Terminalia chebula* belonging to family Combretaceae.	1. Hydrolyzable tannins (20–50%)
	2. Anti-bacterial	Taxonomy	2. Flavonol
	3. Anti-diabetic	Kingdom: Plantae	3. Tannin
	4. Molluscicidal activity	Subkingdom: Trachebionata	4. Triterpenoids
	5. Anti-amoebic and immunomodulatory activity	Division: Magnoliophyte	5. Gallic acid (chebulin)
	6. Anthelminthic	Class: Magnoliopsida	6. Methyl gallate
	7. Wound healing activity	Order: Myrtales	7. Ethyl gallate
	8. Anti-oxidant activity	Genus: Terminalia	8. Chebulagic acid
	8. Anti-anaphylactic activity	Species: Chebula	9. Tetra-O-galloyl -β-D-glucose
Terminalia belerica	1. Anti-oxidant, anti-microbial, and toxicity studies	It consists of seed, fruit, stem bark, and a leaf of *Terminalia belerica* Roxb, belonging to family Combretaceae	1. Glucoside (bellericanin)
	2. Analgesic activity	It is a tree up to 40 m high.	2. Gallotannic acid
	3. Anti-diarrheal activities	The flowers are greenish yellowing color.	3. Allergic acid

Plant	Uses	Biological source	Chemical constituents
	4. Anti-hypertensive studies	It is used in adulteration of *T. arjuna*	4. Gallic acid
	5. Anti-spasmodic and bronchodilators activity	Leaves are petiolate, broadly elliptic, clustered towards the end of branches.	5. Lignans (termilignan and thane lignan)
	6. Hepatoprotective activity		6. 7-hydroxy-3,4-methylenedeoxy
			7. Mannitol
			8. Fructose
			9. Rhamnose
			10. β-sitosterol
Cassia fistula	1. Anti-diabetic activity	It consists of flowers, leaves, bark, the root of *Cassia fistula* Linn belonging to family Caesalpinaceae.	1. Anthraquinones
	2. Anti-hyperlipidemic activity	Taxonomy:	1. Flavonoids
	3. Anti-oxidant activity	Kingdom – Plantae	2. Alkaloids
	4. Anti-pyretic activity	Subkingdom -Trachebinota	3. Terpenoids
	5. Hepatoprotective activity	Superdivision -Spermatophyta	4. Reducing sugar
	6. Anti-inflammatory activity	Division – Mangoliophyta	5. Tannins
	7. Anti-leishmanial activity	Class – Magnoliopsida	6. Carbonyl phlobatannin and steroids
	8. CNS activity	Subclass – Rosidae	7. Fistulic acid
	9. Anti-tumor activity	Order – Fables	8. Rheinglucoside
	10. Anti-microbial activity	Family – Fabacae	9. Galactomannan
	11. Anti-parasitic activity	Genus – Cassia	10. Oxyanthraquinone

TABLE 5.1 *(Continued)*

Plant	Uses	Biological source	Chemical constituents
	12. Wound healing property	Species -Fistula	11. Sennosides A
	13. Anti-ulcer activity		12. Sennosides B
	14. Laxative (main action)		Barbaloin
Bauhinia variegata	1. Anti-diabetic activity	It consists of leaves, bark, seed, flowers of *Bahinia purpurea* Linn belonging to family Leguminosae.	1. β-sitosterol
	2. Anti-inflammatory activity		2. Kaemferol-3-glucoside
	3. Immunomodulatory activity		3. Rutin
	4. Hemaglutinating activity		4. Tannins
	5. Hematinic activity		5. Carbohydrates
	6. Anti-bacterial activity		6. Amides
	7. Hepatoprotective activity		7. Reducing sugars
	8. Anti-bacterial activity		8. Vitamin C
	9. Anti-microbial activity		9. Apigenin 7-O-glucoside
	10. Anti-ulcer		10. Hepatatriacontan-12,13 diol
	11. Anti-carcinogenic		11. Dotetracontan-15-en-9-ol
			12. Protein
			13. Fibers
Mesua ferrea	1. Anti-oxidant and Hepatoprotective activity	It consists of fruit, leaves, flower, seeds of *Mesa ferrea* Linn belonging to family Calophyllaceae.	1. Mesuaxanthone-A

Plant	Uses	Biological source	Chemical constituents
	2. Analgesic activity	Taxonomy	2. Mesuaxanthone-B
	3. Anti-spasmodic activity	Kingdom: Plantae	3. 1,5 dihydroxyanthone (II)
	4. Anti-venom activity	Phylum: Tracheophyta	4. Euxanthone-methyl ether
	5. Anti-neoplastic activity	Order: Malpighiales	5. β-sitosterol
	6. Immunomodulatory activity	Genus: Mesua	6. Ferrol A
	7. Anti-convulsant activity	Height: 18–30 m	7. α-bisabolene
	8. Anti-inflammatory activity	Bark color: Reddish-brown to grey	8. Selinene
	9. Anti-microbial activity	Leaves: simple, acute, lanceolate, and leathery	9. Mesuferrol A
	10. Anti-ulcer activity	Fruits: ovoid with a conical point	10. Mesuferrol B
	11. Anti-arthritic activity		11. Calophyllin-B
			12. Dehydro-cycloguanandin
			13. Euxanthone
			14. Mesuaxanthone-A
			15. Mesuaxanthone-B
			16. Amyrin (α & β)
			17. Lupeol
			18. Caloxanthone C
			19. Betulic acid
Myrtle	1. Anti-inflammatory	Myrtle is scientifically known as *Myrtus communis*, is a species of flowering plant, belonging to the Family Myrtaceae.	a) Leaves contain volatile oils (0.1–0.5%), gallotannins, tannins, myrtucommulone A, and myrtucommulone B

TABLE 5.1 *(Continued)*

Plant	Uses	Biological source	Chemical constituents
	2. Analgesic		b) Essential oils: 1,8-cineole, α-pinene
	3. Anti-bleeding effects		c) Flavonoids- myricetin-3-O-rhamnoside, quercetin-3-O-glucoside, myricetin-3-O-arabinoside
	4. Reduce pain and swelling		d) Other constituent- gallic, caffeic, p-hydroxybenzoic, chlorogenic, linoleic acid, palmitic and oleic acid
Thyme	1. Anti-inflammatory	*Thymus vulgaris* L. is a popular aromatic herb indigenous to the Mediterranean region, belonging to the Family: Labiateae.	a) Non phenolic components (linalool 11.8%, p-cymene 3.2%, alpha pinene 2.26%, α-terpineol 1.12%)
	2. Reduces pain and swelling		b) Phenolic components (carvacrol 57.3%, thymol 10.7%)
	3. Analgesic activity		c) Oxygen-containing monoterpenoids- constitute volatile oil 88.81%
Leek	1. Reduce inflammation	The leek is a vegetable of, *Allium ampeloprasum Subsp iranicum* belonging to the family Amaryllidaceae.	The most active constituents are flavonoids and saponins.
	2. Decreases venostatis and edema		
	3. Decreases sympathomatic recurrence in chronic piles.		

Plant	Uses	Biological source	Chemical constituents
	4. Anti-ulcerogenic activities 5. Reduces the myeloperoxidase activity		
Chinaberry	1. Hypochole-sterolemic	Anti-inflammatory	Flavonoids, terpenoids, anthraquinones
	Cancer protective	It is a deciduous tree or shrub of *Melia azedarach*, belonging to the Family Meliaceae.	Leaf extract contains kaemferol, quercetin, stigmasterol, campsterol, β-sitosterol, pentadecanoic acid, β-carotene, and tocopherol (vitamin E)
	Nematicide		Seeds include β-sitosterol, vanillin, benzoic acid, β-D-glucopyranose, liminoid glycoside viz 1,5-diene-3-O-β-D-glucopyranoside
	Insecticide		Stem bark contains liminoids (7α-acetoxy-14β). Anthraquinone such as 1,3,5,8-tetrahydroxy 2-methylan-thraquinone.
	Hepatoprotective		Root bark contains terpinoids (Azedarachtin-A, Azedarachtin-B)
	Anti-histaminic		Other constituents are quercetin (16.47%), phytol (11.04%), and palmitic acid (15.49%)

5.5 MARKETED POLYHERBAL FORMULATION [30–39]

Arshkeyt™ – A 7-day kit, consist of oral tablets, powders along with topical creams.

Key Ingredients in Tablet

1) Triphala – Analgesic, anti-inflammatory, and wound healing properties.
2) *Amorphallus campanulatus* – Analgesic, anti-bacterial, and anti-oxidant properties.
3) *Melia azedarach* – Analgesic, anti-inflammatory, and anti-microbial activities.

Other Constituent of the Tablets

a) *Cynodon dactylon* is also known as 'Dabashamak' (subsides burning sensation)
b) 'Raktasravbara' (stops bleeding)
c) *Mimosa pudica*, 'sheet virya' (coolant)
d) *Mesua ferrea*, 'Nagakesara' possesses anti-inflammatory, anti-ulcer, and anti-microbial properties.
e) *Baubinia variegata*, 'Vranaropak' (anti-microbial property)
f) *Pengamia pinnata*, 'Arsbabar'- anti-microbial and anti-inflammatory activities.

Key Ingredients in Powder

1) Isabgul (*Plantago ovata*): High fibre bulk-forming laxative to reduce the bleeding episodes in case of internal bleeding.
2) Cassia fistula: 'Shoolhara' (pain killer) and 'Stramsan' (mild laxative); possess anti-inflammatory and anti-microbial properties.
3) *G. glabra*: wound healing.
4) Fennel: *Foeniculum vulgare*, belonging to family Umbelliferae; possess anti-inflammatory, anti-microbial, and laxative treatment.

Key Ingredients of Cream

1) Oil *P. pinnata* and oil *Azadirachta indica* possess anti-microbial and anti-inflammatory properties.

2) *Vateria indica, Berberisa ristala, C. doctylon, Curcuma longa,* and *M. pudica* described as 'Vranaropak' – stops bleeding.
3) Yasbada Bhasma, comprising of zinc and borax; used as a disinfectant improves binding powers of a cell, soft tissues, improves cell migration, and cell regeneration.
4) *Hemidesmus indicus* and *Vitex negendo* described as 'Shothbara' (subsides swelling) and 'Shoolbara' (pain killer) exhibit anti-inflammatory, analgesics, and anti-microbial activity.

5.6 CONCLUSION

In traditional medicine, these drugs were used as a remedy to treat hemorrhoids. But still, there are many drugs that are yet to be explored for anti-hemorrhoidal activity. The ethnopharmacological traditional knowledge of several plants with potential anti-hemorrhoid activities like *Balsamodendron mukkul*, Shilajit, *Emblica officinalis, Berberis aristate, Terminalia chebula, Cassia Fistula, Bauhinia variegate, Mesua ferrea, Thymus vulgaris, Allium ampeloprasum, Melia azedarach, Myrtle communis*, etc. opened new doors for clinical practices and further research.

KEYWORDS

- **hemorrhoids**
- **herbal treatment**
- **inflammation**
- **pathogenesis**
- **preventive measures**
- **rectal bleeding**
- **risk factors**

REFERENCES

1. Lohsiriwat, V., (2012). Hemorrhoids: From basic pathophysiology to clinical management. *World Journal of Gastroenterology: WJG, 18*(17), 2009–2017.

2. Sun, Z., & Migaly, J., (2016). Review of hemorrhoid disease: Presentation and management. *Clinics in Colon and Rectal Surgery, 29*(1), 022–029.
3. Sandler, R. S., & Peery, A. F., (2019). Rethinking what we know about hemorrhoids. *Clinical Gastroenterology and Hepatology, 17*(1), 8–15.
4. Gibbons, C. P., Bannister, J. J., & Read, N. W., (1988). Role of constipation and anal hypertonia in the pathogenesis of hemorrhoids. *British Journal of Surgery, 75*(7), 656–660.
5. Haas, P. A., Fox, T. A., & Haas, G. P., (1984). The pathogenesis of hemorrhoids. *Diseases of the Colon and Rectum, 27*(7), 442–450.
6. Kranthi, G., Krishna, M. C. H., Kali, T. S., & Mahendra, K. T., (2010). Herbal Treatment for Hemorrhoids. *Journal of Innovative Trends in Pharmaceutical Sciences, 1*(5), 236–244.
7. Vastrad, C. S., & Pakkanavar, R. V., (2002) . Clinical evaluation of PIL-28, a herbal formulation in the management of hemorrhoids. *The Antiseptic, 9*(99), 343–344.
8. Sharma, S., Khan, N., & Sultana, S., (2005). Balsamodendron mukul suppresses benzoyl peroxide and ultraviolet light induced tumor promotional events in Swiss mice. *Journal of Photochemistry and Photobiology B: Biology, 78*(1), 43–51.
9. Agarwal, S. P., Khanna, R., Karmarkar, R., Anwer, M. K., Khar, R. K., Shilajit., (2007). *A Review. Phytother Res., 21*(5), 401–405.
10. Agarwal, S. P., Khanna, R., Karmarkar, R., Anwer, M. K., & Khar, R. K., (2007). Shilajit: A review. *Phytotherapy Research: An International Journal Devoted to Pharmacological and Toxicological Evaluation of Natural Product Derivatives, 21*(5), 401–405.
11. Ervina, M., (2018). A review: *Melia azedarach* L. as a potent anticancer drug. *Pharmacognosy Reviews, 12*(23), 94-102.
12. Shahid, M., Rahim, T., Shahzad, A., Latif, T. A., Fatma, T., Rashid, M., Raza, A., & Mustafa, S., (2009). Ethnobotanical studies on *Berberis aristata* DC. root extracts. *African Journal of Biotechnology, 8*(4) 556-563.
13. Potdar, D., Hirwani, R. R., & Dhulap, S., (2012) . Phyto-chemical and pharmacological applications of *Berberis aristata. Fitoterapia, 83*(5), 817–830.
14. Ram, M. S., Neetu, D., Yogesh, B. L., Anju, B., Dipti, P., Pauline, T., Sharma, S. K., et al., (2002) . Cyto-protective and immunomodulating properties of amla (*Emblica officinalis*) on lymphocytes: An *in-vitro* study. *Journal of Ethnopharmacology, 81*(1), 5–10.
15. Jain, R., Pandey, R., Mahant, R. N., & Rathore, D. S., (2015). A review on medicinal importance of emblica officinalis. *International Journal of Pharmaceutical Sciences and Research, 6*(1), 72–84.
16. Basha, S. J., & Code, Q. R., (2017). A review on *Terminalia chebula. Int. J. Pharm. Res., 7*(10), 187–191.
17. Motamarri, S. N., Karthikeyan, M., Kannan, M., & Rajasekar, S., (2012) . *Terminalia belerica* roxb: A phytopharmacological review. *International Journal of Research in Pharmaceutical and Biomedical Sciences, 3*, 96–99.
18. Kumar, N., & Khurana, S. M., (2018). Phytochemistry and medicinal potential of the *Terminalia bellirica* roxb.(Bahera). *Indian Journal of Natural Products and Resources (IJNPR) [Formerly Natural Product Radiance (NPR)], 9*(2) , 97–107.
19. Ira, I., Hossain, M. M., Zannat, N. N., Uddin, A. M., & Ali, M. A., (2014). *Cassia fistula* linn: A review of phytochemical and pharmacological studies. *International Journal of Pharmaceutical Sciences and Research, 5*(7. , 2125–2130.
20. Shahana, S., Nikalje, A. P., (2017). *A Brief Review on Bauhinia variegata: Phytochemistry, Antidiabetic and Antioxidant potential. Am. J. Pharmtech Res., 7*, 186-97.

21. Sahu, G., & Gupta, P. K., (2012) . A review on *Bauhinia variegata* Linn. *International research Journal of Pharmacy, 3*(1), 48–51.
22. Kumar, T., & Chandrashekar, K. S., (2011). *Bauhinia purpurea* Linn.: A review of its ethnobotany, phytochemical and pharmacological profile. *Res. J. Med. Plant, 5*(4), 420–431.
23. Sharma, A., Sharma, S., & Parashar, B., (2017). *Mesua ferrae* Linn: A review of the Indian medical herb. *Systematic Reviews in Pharmacy, 8*(1), 19-23.
24. Chahar, M. K., DS, S. K., Geetha, L., Lokesh, T., & Manohara, K. P., (2013). *Mesua ferrea* L.: A review of the medical evidence for its phytochemistry and pharmacological actions. *African Journal of Pharmacy and Pharmacology, 7*(6), 211–219.
25. Asif, M., Jafari, S. F., Iqbal, Z., Revadigar, V., Oon, C. E., Abdul, A. S., & Majid, A. M., (2017). Ethnobotanical and Phyto-pharmacological attributes of *Mesua ferrea*: A mini review [J]. *J Appl Pharml Sci., 7*(4), 242–251.
26. Mahboubi, M., (2017). Effectiveness of *Myrtus* communis in the treatment of hemorrhoids. *Journal of Integrative Medicine, 15*(5), 351–358.
27. Kumar, A., Shukla, R., Singh, P., Prasad, C. S., & Dubey, N. K., (2008). Assessment of *Thymus vulgaris* L. essential oil as a safe botanical preservative against post-harvest fungal infestation of food commodities. *Innovative Food Science and Emerging Technologies, 9*(4), 575–580.
28. Ho, Y. H., Foo, C. L., Seow-Choen, F., & Goh, H. S., (1995). Prospective randomized controlled trial of a micronized flavonidic fraction to reduce bleeding after haemorrhoidectomy. *British Journal of Surgery, 82*(8. , 1034–1035.
29. Sultana, S., Asif, H. M., Akhtar, N., Waqas, M., & Rehman, S. U., (2014). Comprehensive review on ethanobotanical uses, phytochemistry and pharmacological properties of *Melia azedarach* Linn. *Asian Journal of Pharmaceutical Research and Health Care, 6*(1), 26-32.
30. Shilpi, J. A., Ray, P. K., Sarder, M. M., & Uddin, S. J., (2005). Analgesic activity of *Amorphophallus* campanulatus tuber. *Fitoterapia, 76*(3, 4), 367–369.
31. Zhang, W. M., Liu, J. Q., Peng, X. R., Wan, L. S., Zhang, Z. R., Li, Z. R., & Qiu, M. H., (2014). Triterpenoids and sterols from the leaves and twigs of *Melia azedarach. Natural Products and Bioprospecting, 4*(3), 157–162.
32. Biswas, T. K., & Mukherjee, B., (2003). Plant medicines of Indian origin for wound healing activity: A review. *The International Journal of Lower Extremity Wounds, 2*(1), 25–39.
33. Kokane, D. D., More, R. Y., Kale, M. B., Nehete, M. N., Mehendale, P. C., & Gadgoli, C. H., (2009). Evaluation of wound healing activity of root of *Mimosa pudica. Journal of Ethnopharmacology, 124*(2) , 311–315.
34. Mazumder, R., Dastidar, S. G., Basu, S. P., Mazumder, A., & Kumar, S., (2003). Emergence of *Mesua ferrea* Linn. leaf extract as a potent bactericide. *Ancient Science of Life, 22*(4), 160-165.
35. Mishra, A., Sharma, A. K., Kumar, S., Saxena, A. K., & Pandey, A. K., (2013). *Bauhinia variegata* leaf extracts exhibit considerable antibacterial, antioxidant, and anticancer activities. *BioMed Research International, 2013.*
36. Al Muqarrabun, L. M., Ahmat, N., Ruzaina, S. A., Ismail, N. H., & Sahidin, I., (2013). Medicinal uses, phytochemistry and pharmacology of *Pongamia pinnata* (L.) Pierre: A review. *Journal of Ethnopharmacology, 150*(2) , 395–420.

37. Tripathi, R. K., Bolegave, S. S., Shetty, P. A., Uchil, D. A., Rege, N. N., Chawda, M. B., & Rege, S. A., (2015). Efficacy and safety of a polyherbal formulation in hemorrhoids. *Journal of Ayurveda and Integrative Medicine, 6*(4), 225-232.
38. Bhalodia, N. R., & Shukla, V. J., (2011). Antibacterial and antifungal activities from leaf extracts of *Cassia fistula* L.: An ethno medicinal plant. *Journal of Advanced Pharmaceutical Technology and Research., 2*(2) , 104-109.
39. Das, S., & Singh, B. S., (2013). The bioactive and therapeutic potential of *Hemidesmus indicus* R. Br. (*Indian sarsaparilla*) root. *Phytotherapy Research, 27*(6), 791–801.

CHAPTER 6

Precision, Immunotherapy, Immuno Oncology, and Research in Cancer

FRANCISCO TORRENS[1] and GLORIA CASTELLANO[2]

[1]*Institut Universitari de Ciència Molecular, Universitat de València, Edifici d'Instituts de Paterna, P.O. Box 22085, E-46071 València, Spain, E-mail: torrens@uv.es*

[2]*Departamento de Ciencias Experimentales y Matemáticas, Facultad de Veterinaria y Ciencias Experimentales, Universidad Católica de Valencia San Vicente Mártir, Guillem de Castro-94, E-46001 València, Spain*

ABSTRACT

Cancer is not a disease but a series of diseases. Without pre-clinical and clinical research, it would be difficult to continue following. Nivolumab + Ipilimumab provides a better overall survival for melanoma. Prostate cancer SPARTAN shows a trend to benefit. In endometrial cancer, IMvigor130 tested chemo + immunotherapy in patients eligible/ineligible for chemotherapy but today is not a standard; NABUCCO neoadjuvant Ipilimumab/Nivolumab was tolerated in high-risk urothelial carcinoma. In breast cancer, MONALEESA-3/7 data represent the largest evidence for survival for cyclin-dependent-kinases inhibitor without a match. Ribociclib shows significant benefit in overall survival in MONALEESA-3/7. Abemaciclib shows significant benefit in overall survival in MONARCH 2. CDICAB inhibitors are selection drugs in advanced-luminal-treatment first/second lines. The Nivolumab + Ipilimumab confirms a strong activity in cervical cancer. Pembrolizumab + Lenvatinib show excellent results in endometrial cancer. Ovarian cancer first-line Veliparib shows excellent results with a rise in progression-free survival in all patients. Surgically ablated patients are excluded. The primary objective is met. Results are got in BRCA mutation and homologous recombination deficiency subgroups. Central arm exists with relative volume change. In lung cancer internal survival, Atezolizumab increases survival *vs.* chemotherapy PDL1/TC3/IC3. Nivolumab + Ipilimumab improves overall survival *vs.* chemotherapy.

6.1 INTRODUCTION

Setting the scene: oncology in 21st century, precision medicine (PM) *vs.* immunotherapy (IT), immuno-oncology drug development, how to do more with less, the cancer problem closer to the cure, translational clinical research, melanoma, genitourinary (GUC), breast (BRCA, BC), gynecological, lung (LC), head and neck (HNC), digestive (DC), and miscellaneous (MC) cancers.

Cancer is not a disease but a series of diseases. Without pre-clinical and clinical research, it would be difficult to continue following. Nivolumab + Ipilimumab (Nivo/Ipi) provides a better overall survival (OS) for melanoma. Prostate cancer (PC) SPARTAN study shows a trend to benefit. In endometrial cancer (EC), IMvigor130 tested CT + IT in patients eligible/ineligible for chemotherapy (CT) but today is not a standard; NABUCCO neoadjuvant Ipi/Nivo were tolerated in high-risk urothelial carcinoma. In BC, MONALEESA-3/7 data represent the largest evidence for OS for cyclin-dependent-kinases (CDKs) inhibitor without a match. Ribociclib shows significant benefit in OS in MONALEESA-3 and -7. Abemaciclib shows significant benefits in OS in MONARCH 2 study. CDICAB inhibitors are selection drugs in advanced-luminal-treatment first/second lines (1/2L). The Nivo + Ipi combination confirms a strong activity in cervical cancer (CeC). Pembrolizumab + Lenvatinib combination shows excellent results in EC. Ovarian cancer (OC) 1L Veliparib shows excellent results with a rise in progression-free survival (PFS) in all patients. Surgically ablated patients are excluded. The primary objective is met. Results are got in BRCA mutation (mBRCA) and homologous recombination deficiency (HRD) subgroups. There is a central arm with relative volume change. In LC internal OS, Atezolizumab increases OS *vs.* CT PDL1/TC3/IC3. Nivo + Ipi improves OS *vs.* CT.

In earlier publications it was informed the modeling of complex multicellular systems and tumor–immune cells competition [1], information-theoretic entropy for molecular classification of oxadiazolamines as potential therapeutic agents [2], molecular classification of 5-amino-2-aroylquinolines and 4-aroyl-6,7,8-trimethoxyquinolines as highly potent tubulin polymerization inhibitors [3], polyphenolic phytochemicals in cancer prevention, therapy, bioavailability *vs.* bioefficacy [4], molecular classification of anti-tubulin agents with indole ring binding at colchicine-binding site [5], molecular classification of 2-phenylindole-3-carbaldehydes as potential antimitotic agents in human BC cells [6], cancer, its hypotheses [7], personalized PM (PPM) from theory to practice and cancer [8]. It was reported how human immunodeficiency virus/acquired immunodeficiency syndrome destroy immune defenses, hypothesis [9], 2014 emergence, spread, uncontrolled Ebola outbreak [10, 11], Ebola virus disease, questions, ideas, hypotheses, models [12], swine, avian, human influenza, origins of A (H1N1) influenza pandemic strain [13], clinical translational research, cancer, diabetes, cardio-vascular disease [14] epigenomics, epithelial plasticity, clinical genetics and rare diseases [15]. The present report reviews oncology in the 21st century, PM *vs.* IT, immuno-oncology drug development, how to do more with less, the cancer problem closer to the cure, translational clinical research, melanoma, GUC, BC, gynecological, LC, HNC, DC, and MC. This work aims to initiate a debate by suggesting several questions (Q), which can arise when addressing subjects of translational chemical research on oncology, and providing, when possible, answers (A) and hypotheses (H).

6.2 ONCOLOGY IN 21ST CENTURY: PRECISION MEDICINE *VS.* IMMUNOTHERAPY

Barbacid Montalbán reviewed an H on oncology in the 21st century and PM *vs.* IT [16].

H1. *NTRK* fusion-positive cancers and TRK inhibitor therapy [17].

He provided the following conclusions (Cs).

C1. Cancer is not a disease but a series of diseases.
C2. Without pre-clinical as well as clinical research, it would be difficult to continue following.

6.3 IMMUNO-ONCOLOGY DRUG DEVELOPMENT: DO MORE WITH LESS

Cancer treatment was revolutionized *via* drugs that harness the power of the patient's immune system (IS) [18]. The development and characterization of the blockbuster drugs require a series of complex cell-based assays to analyze the effect on different facets of IS.

6.4 THE CANCER PROBLEM: CLOSER TO THE CURE

Cancer is one of the deadliest illnesses humanity knew [19]. Scientists and physicians labored for decades to find a cure, but it is important to accept that a cure may be far more complex than thought.

6.5 II DAY ON TRANSLATIONAL CLINICAL RESEARCH POST-ESMO 2019

Lluch Hernández organized II Day on *Translational Clinical Research Post-ESMO 2019* [20].

6.6 ROUND TABLE 1: MELANOMA/GENITOURINARY/BREAST/ GYNAECOLOGICAL CANCERS

Torregrosa Maicas and Soriano Teruel moderated Round Table (RoT)1 on melanoma, GUC, BC and gynaecological cancers [21]. González Barallo proposed H/Q on melanoma [22].

H1. Localized high-risk (HR) disease.
H2. Primary endpoint: relapse-free survival (RSF) in all patients.
H3. Biomarker (MB) analyses.
Q1. What do people know?

She raised the following open questions.

Q2. Have people reached the plateau of the RFS curve?
Q3. This benefit, is it going to translate into OS?

She provided the following conclusion.

C1. Nivo + Ipi provides a better OS.

In GUC, Juan Fita raised the following question on PC [23].

Q4. Not the primary endpoint but wonderful results; what to do?

She provided the following conclusion.

C2. SPARTAN study shows a trend to benefit.

She provided an additional conclusion on EC.

C3. IMvigor130 is the first trial to test CT + IT in patients eligible/ineligible for CT, but today is not a standard; NABUCCO neoadjuvant Ipi/Nivo were tolerated in HR urothelial carcinoma.

Hernando Meliá provided the following conclusions on BC [24].

C4. The combined dataset of MONALEESA-3/7 studies represents 1,400 patients and the largest body of evidence for OS for any CDK4/6 inhibitor, and the data demonstrate a did not match with any results.

C5. Ribociclib shows significant benefit in OS in MONALEESA-3 and MONALEESA-7 studies.

C6. Abemaciclib shows significant benefits in OS in MONARCH 2 study.

She raised the following question.

Q5. Does triple-negative BC (TNBC) benefit?

She provided an additional conclusion.

C7. CDICAB inhibitors are the selection of drugs in 1/2L of advanced luminal treatment.
In GUC, Iranzo Sabater provided the following conclusion on CeC [25].

C8. The combination of Nivo/Ipi confirms a strong activity in this type of tumor.

She proposed the following hypothesis on EC.

H4. Synergistic effect of Pembrolizumab and Lenvatinib.

She provided an additional conclusion.

C9. Pembrolizumab + Lenvatinib combination shows excellent results.

She provided additional conclusions on OC.

C10. Excellent results in 1L Veliparib with a significant rise in PFS in all patients.
C11. Surgically ablated R0 patients are excluded.
C12. The primary objective is met.
C13. Good results are obtained in mBRCA and HRD subgroups.
C14. There is a central arm with relative volume change (RVC).

She proposed additional questions and hypotheses.

Q5. Poly [adenosine diphosphate (ADP)-ribose] polymerase (PARP) inhibitors (iPARPs) in 1L for all?
Q6. In addition, in relapse?
H5. The *dogma* of the residual population.

6.7 ROUND TABLE 2: LUNG/HEAD AND NECK/DIGESTIVE/ MISCELLANEOUS CANCERS

Juan Vidal and Blasco Cordellat moderated RoT2 on LC, HNC, DC and MC [26]. Blasco Mollá proposed an H on LC [27].
H1. IT in 1L: objective Ipi *vs*. CT.

She provided the following conclusions.

C1. Internal OS analyses: Atezolizumab increases OS *vs*. CT programmed death-ligand (PDL)1/TC3/IC3.
C2. Nivo/Ipi improves OS *vs*. CT.

She raised the following question.

Q1. With what level of PDL1?

She provided an additional conclusion.

C3. It seems that tissue tumor mutational burden (tTMB) is not a predictive factor.
She proposed an additional hypothesis.

H2. tTMB exploratory analysis as a predictive factor.

She provided additional conclusions.

C4. One cannot draw conclusions.
C5. Different techniques and dissimilar methods.

She proposed an additional hypothesis.

H3. FLAURA double-blind study design.

She provided an additional conclusion.

C6. Osimertinib, robust data in OS. Less toxicity. It sits on 1L.

Ochenduszko provided the following conclusion on HNC [28].

C7. Radiotherapy (RT) plus the primary treatment and lymph node (LN) added to CT significantly improve OS in CT.

He proposed the following questions and hypothesis.

Q2. Utility?
Q3. How many patients?
Q4. Reproducibility in our region?
H4. Geriatric assessment (GA): Personalized treatment according to GA in 1L recurrent HNC.

He provided an additional conclusion on GA.

C8. ELAN-FIT trial presents two buts (Bs).
B1. Is the EXTREME scheme the standard treatment?
B2. In addition, Paclitaxel +/– Cetuximab?

He provided an additional conclusion.

C9. Radical RT of the primary tumor or metastatic nasopharyngeal carcinoma.

Díaz Beveridge raised the following questions on DC [29].

Q5. Neoadjuvant CT: What do we learn from FOxTROT study?
Q6. How accurate is CT staging to define Stage II or Stage III?

He proposed the following Hs to be further validated [microsatellite instability/deficient mismatch repair (MSI/dMMR)].

H5. Chemo-resistant.
H6. Treatment with checkpoint inhibitors.

He raised an additional question.

Q7. Downsizing/downstaging?

He proposed the following conclusions.

C10. There is a doubt about downsizing/downstaging.
C11. Reduced resection margin R1 reset.

He raised additional questions on the residual disease.

Q8. Can circulating tumor deoxyribonucleic acid (ctDNA) be used to detect minimal residual disease (MRD)?
Q9. Is ctDNA predictive of recurrence?

He proposed additional conclusions.

C12. It confirms previous results that ctDNA is a powerful predictor of relapse.
C13. Mutation tracking increases MRD and predictive accuracy.

He raised an additional question.

Q10. Is ctDNA predictive for treatment duration?

He proposed an additional conclusion on recurrent advanced cancer of the colon (CC).

C14. In the BEACON-colorectal cancer (CRC) study, Encorafenib was used.

He raised additional questions on advanced hepatocarcinoma.

Q11. The future of the future?
Q12. Are there combinations of therapies?

He proposed additional conclusions.

C15. Role of neoadjuvant CT in localized CC.
C16. Role of liquid biopsy (LB) in the management of localized CC.
C17. New treatment options in CC: BRAF.
Q13. (Ana Lluch Hernández). How to treat patients with ctDNA even without a clinical action?
A13. This is for those patients who have an active load and metastatic activity

Garde Noguera presented MC [30]. He proposed a conclusion on pancreatic cancer.

C18. NEOLAP is the first prospective, randomized study to compare two different induction CT regimens (Nab-Paclitaxel/Gemcitabine *vs.* FOLFIRINOX-based) in that setting.

He proposed an additional conclusion on cholangiocarcinoma.

C19. A total of 56 unique FGFR2 fusion genes were observed.

He proposed an additional conclusion on mesothelioma.

C20. ETOP PROMISE-meso is the first work to show a positive impact on mesothelioma.

He raised the following question on glioblastoma.

Q14. Regorafenib?

He proposed an additional conclusion on glioblastoma.

C21. Regorafenib does not harm the patient's quality of life.

6.8 FINAL REMARKS

From the present results and discussion, the following final remarks can be drawn.

1. Cancer is not a disease but a series of diseases.
2. Without pre-clinical and clinical research, it would be difficult to continue following.
3. Nivolumab + Ipilimumab provides a better overall survival for melanoma.
4. Prostate cancer SPARTAN study shows a trend to benefit.
5. In endometrial cancer, IMvigor130 tested chemo + immunotherapy in patients in/eligible for chemotherapy but today is not a standard; NABUCCO neoadjuvant Ipilimumab/Nivolumab was tolerated in high-risk urothelial carcinoma.
6. In breast cancer, MONALEESA-3/7 data represent the largest evidence for survival for cyclin-dependent-kinases inhibitor without a match.
7. Ribociclib shows significant benefit in overall survival in MONALEESA-3 and -7.
8. Abemaciclib shows significant benefits in overall survival in MONARCH 2 study.
9. CDICAB inhibitors are selection drugs in advanced-luminal-treatment first/second lines.
10. The Nivolumab + Ipilimumab combination confirms a strong activity in cervical cancer.
11. Pembrolizumab + Lenvatinib combination shows excellent results in endometrial cancer.
12. Ovarian cancer first-line Veliparib shows results with a rise in progression-free survival.
13. Surgically ablated patients are excluded.
14. The primary objective is met.
15. Results are got in BRCA mutation/homologous recombination deficiency subgroups.
16. There is a central arm with relative volume change.
17. In lung cancer internal survival, Atezolizumab increases survival *vs.* chemotherapy PDL1/TC3/IC3.
18. Nivolumab + Ipilimumab improves overall survival *vs.* chemotherapy.
19. It seems that tissue tumor mutational burden is not a predictive factor.
20. One cannot draw conclusions.
21. Different techniques and dissimilar methods.
22. Osimertinib, robust data in overall survival. Less toxicity. It sits on the first line.

23. In head/neck cancer, radiotherapy plus the primary treatment/lymph node added to chemotherapy improve survival in chemotherapy.
24. In the geriatric assessment, the ELAN-FIT trial presents two buts.
25. Radical radiotherapy of the primary tumor or metastatic nasopharyngeal carcinoma.
26. In digestive cancer, there is a doubt about downsizing/downstaging.
27. Reduced resection margin reset.
28. In residual disease, circulating tumor deoxyribonucleic acid is a relapse predictor.
29. Mutation tracking increases minimal residual disease and predictive accuracy.
30. In the BEACON-colorectal cancer study, Encorafenib was used.
31. Role of neoadjuvant chemotherapy in localized colon cancer.
32. Role of liquid biopsy in the management of localized colon cancer.
33. New treatment options in colon cancer: BRAF.
34. In pancreatic cancer, NEOLAP compares two induction chemotherapy regimens.
35. A total of 56 unique FGFR2 fusion genes were observed in cholangiocarcinoma.
36. ETOP PROMISE-meso is the first work to show a positive impact on mesothelioma.
37. Regorafenib does not harm the glioblastoma patient's quality of life.

ACKNOWLEDGMENTS

The authors thanks the support from Fundacion Universidad Catolica de Valencia San Vicente Martir (Project No. 2019-217-001UCV).

KEYWORDS

- **cancer cure**
- **cancer problem**
- **immuno-oncology drug development**
- **melanoma**
- **oncology**
- **precision medicine**
- **translational clinical research**

REFERENCES

1. Torrens, F., & Castellano, G., (2009). modeling of complex multicellular systems: Tumor–immune cells competition. *Chem. Central J., 3*(Suppl. I), 75–1–1.
2. Torrens, F., & Castellano, G., (2013). Information theoretic entropy for molecular classification: Oxadiazolamines as potential therapeutic agents. *Curr. Comput.-Aided Drug Des., 9*, 241–253.
3. Torrens, F., & Castellano, G., (2013). Molecular classification of 5-amino-2-aroylquinolines and 4-aroyl-6,7,8-trimethoxyquinolines as highly potent tubulin polymerization inhibitors. *Int. J. Chemoinf. Chem. Eng., 3*(2), 1–26.
4. Estrela, J. M., Mena, S., Obrador, E., Benlloch, M., Castellano, G., Salvador, R., & Dellinger, R. W., (2017). Polyphenolic phytochemicals in cancer prevention and therapy: Bioavailability versus bioefficacy. *J. Med. Chem., 60*, 9413–9436.
5. Torrens, F., & Castellano, G., (2018). Molecular classification of antitubulin agents with indole ring binding at colchicine-binding site. In: Parikesit, A. A., (ed.), *Molecular Insight of Drug Design* (pp. 47–67). InTechOpen: Vienna.
6. Torrens, F., & Castellano, G., (2019). Molecular classification of 2-phenylindole-3-carbaldehydes as potential antimitotic agents in human breast cancer cells. In: Haghi, A. K., Thomas, S., Praveen, K. M., & Pai, A. R., (eds.), *Theoretical Models and Experimental Approaches in Physical Chemistry: Research Methodology and Practical Methods* (pp. 221–248). Apple Academic–CRC Press: Waretown, NJ.
7. Torrens, F., & Castellano, G., (2020). Cancer and hypotheses on cancer. In: Pogliani, L., Torrens, F., & Haghi, A. K., (eds.), *Molecular Chemistry and Biomolecular Engineering: Integrating Theory and Research with Practice* (pp. 129–248). Apple Academic–CRC Press: Waretown, NJ.
8. Torrens, F., & Castellano, G., (2020). Precision personalized medicine from theory to practice: Cancer. In: Aguilar, C. N., Ameta, S. C., & Haghi, A. K., (eds.), *Green Chemistry and Biodiversity: Principles, Techniques, and Correlations* (pp. 209–242). Apple Academic–CRC Press: Waretown. NJ.
9. Torrens, F., & Castellano, G., (2014) AIDS destroys immune defenses: Hypothesis. *New Front. Chem., 23*, 11–20.
10. Torrens-Zaragozá, F., & Castellano-Estornell, G., (2015). Emergence, spread and uncontrolled Ebola outbreak. *Basic Clin. Pharmacol. Toxicol., 117*(Suppl. 2), 38.
11. Torrens, F., & Castellano, G., (2014). spread/uncontrolled Ebola outbreak. *New Front. Chem.* 2015, *24*, 81–91.
12. Torrens, F., & Castellano, G., (2016). Ebola virus disease: Questions, ideas, hypotheses and models. *Pharmaceuticals, 9*, 14–6–6.
13. Torrens, F., & Castellano, G., (2020). Swine, avian, and human flu, and origins of a (H1N1) flu pandemic strain. In: Torrens, F., Mahapatra, D. K., & Haghi, A. K., (eds.), Physical Biochemistry, Biophysics, and Molecular Chemistry: Applied Research and Interactions (pp. 53–68.). Apple Academic–CRC: Waretown, NJ.
14. Torrens, F., & Castellano, G., (2020). Clinical translational research: Cancer, diabetes, and cardiovascular disease. In: Torrens, F., Mahapatra, D. K., & Haghi, A. K., (eds.), *Physical Biochemistry, Biophysics, and Molecular Chemistry: Applied Research and Interactions* (pp. 35-52). Apple Academic–CRC Press: Waretown, NJ.
15. Torrens, F., & Castellano, G., (2021). Epigenomics, epithelial plasticity, clinical genetics, and rare diseases. In: Khalaf, M. N., Smirnov, M. O., Kannan, P., & Haghi,

A. K., (eds.), *Environmental Technology and Engineering Techniques: Basic Concepts and Health Interventions* (pp. 247–257). Apple Academic–CRC Press: Waretown, NJ.
16. Barbacid, M. M., (2016). *Personal Communication.*
17. Cocco, E., Scaltriti, M., & Drilon, A., (2018). *NTRK* fusion-positive cancers and TRK inhibitor therapy. *Nat. Rev. Clin. Oncol., 15*, 731–747.
18. Chapple, K., & Gomez-Donart, A., (2019). Do more with less. *Eur. Biopharm. Rev., 2019*(88), 46–50.
19. Rommel, C., (2019). Closer to the cure. *Eur. Biopharm. Rev., 2019*(88), 52–56.
20. Lluch, H. A., (2019). Book of Abstract II Post-ESMO 2019 Clinical Translational Research Conference, Institute of Health Research (INCLIVA). Valencia, Spain, INCLIVA, Valencia, Spain, 2019.
21. Torregrosa, M. M., & Soriano, T. V., (2019). Book of Abstract II Post-ESMO 2019 Clinical Translational Research Conference, Institute of Health Research (INCLIVA). Valencia, Spain, INCLIVA, Valencia, Spain, 2019.: RT-1.
22. González, B. I., (2019). Book of Abstract II Post-ESMO 2019 Clinical Translational Research Conference, Institute of Health Research (INCLIVA). Valencia, Spain, INCLIVA, Valencia, Spain, 2019: RT-1.
23. Juan, F. M. J., (2019). Book of Abstract II Post-ESMO 2019 Clinical Translational Research Conference, Institute of Health Research (INCLIVA). Valencia, Spain, INCLIVA, Valencia, Spain, 2019: RT-1.
24. Hernando, M. C., (2019). Book of Abstract II Post-ESMO 2019 Clinical Translational Research Conference, Institute of Health Research (INCLIVA). Valencia, Spain, INCLIVA, Valencia, Spain, 2019: RT-1.
25. Iranzo, S. A., (2019). Book of Abstract II Post-ESMO 2019 Clinical Translational Research Conference, Institute of Health Research (INCLIVA). Valencia, Spain, INCLIVA, Valencia, Spain, 2019: RT-1.
26. Juan, V. O., & Blasco, C. A., (2019 Book of Abstract II Post-ESMO 2019 Clinical Translational Research Conference, Institute of Health Research (INCLIVA). Valencia, Spain, INCLIVA, Valencia, Spain, 2019: RT-2.
27. Blasco, M. S., (2019). Book of Abstract II Post-ESMO 2019 Clinical Translational Research Conference, Institute of Health Research (INCLIVA). Valencia, Spain, INCLIVA, Valencia, Spain, 2019: RT-2.
28. Ochenduszko, S., (2019). Book of Abstract II Post-ESMO 2019 Clinical Translational Research Conference, Institute of Health Research (INCLIVA). Valencia, Spain, INCLIVA, Valencia, Spain, 2019: RT-2.
29. Díaz, B. R., (2019). Book of Abstract II Post-ESMO 2019 Clinical Translational Research Conference, Institute of Health Research (INCLIVA). Valencia, Spain, INCLIVA, Valencia, Spain, 2019: RT-2.
30. Garde, N. J., (2019). Book of Abstract II Post-ESMO 2019 Clinical Translational Research Conference, Institute of Health Research (INCLIVA). Valencia, Spain, INCLIVA, Valencia, Spain, 2019: RT-2.

CHAPTER 7

Oxidative Stress-Protective/Anti-Melanogenic Effects of Loliolide

FRANCISCO TORRENS[1] and GLORIA CASTELLANO[2]

[1]*Institut Universitari de Ciència Molecular, Universitat de València, Edifici d'Instituts de Paterna, P.O. Box 22085, E-46071 València, Spain, E-mail: torrens@uv.es*

[2]*Departamento de Ciencias Experimentales y Matemáticas, Facultad de Veterinaria y Ciencias Experimentales, Universidad Católica de Valencia San Vicente Mártir, Guillem de Castro-94, E-46001 València, Spain*

ABSTRACT

Neuroinflammation is a major contributor to the start and development of degenerative diseases, for example, Alzheimer's and Parkinson's diseases. Microglial and astrocyte activation is a brain defense mechanism to counteract harmful pathogens and damaged tissues, while their prolonged activation induces neuroinflammation that triggers or exacerbates degeneration. No pharmacological therapies exist able to slow down or stop degeneration progression. Research is turning to the identification of natural products with protective action *vs.* diseases. Considering the role of neuroinflammation in the onset and development of degenerative pathologies, natural products with anti-inflammatory activity could be good candidates for therapies. Marine organisms represent an enormous source of natural products and, among them, algae are appreciated sources of important bioactive substances, for example, antioxidants, proteins, vitamins, minerals, soluble dietary fibers, polyunsaturated fatty acids, polysaccharides, sterols, carotenoids, tocopherols, terpenes, phycobilins, phycocolloids, and phycocyanins. Many anti-inflammatory products were isolated from marine algae with potential protective efficacy *vs.* neuroinflammation. Loliolide is a monoterpenoid hydroxylactone found in many algae, for example, freshwater green alga *Prasiola japonica*. It and products in *P. japonica* were not studied systematically with respect to skin pharmacology.

7.1 INTRODUCTION

Setting the scene: Phycotoxin, commercial algal product, uses, phycoremediation, oxidative stress (OS)-protective action (PA) and anti-melanogenic (AM) action (AMA) of monoterpene lactone (MTL) γ-lactone loliolide and ethanol CH_3CH_2OH extract from freshwater green alga (GA) *Prasiola japonica* (Pj-EE), and fucoxanthin-rich brown algae (BAs) extract improving male reproductive function on streptozotocin (SZT)-nicotinamide (NAM)-induced diabetic rat model. Neuroinflammation is the main contributor to the onset and progression of neurodegenerative diseases, for example, Alzheimer's (AD) and Parkinson's (PD) diseases. Microglial and astrocyte activation is a brain defense mechanism to counteract harmful pathogens and damaged tissues, while their prolonged activation induces neuroinflammation that triggers or exacerbates neurodegeneration. There is no pharmacological therapy able to slow down or stop the progression of neurodegeneration. Research is turning to the identification of natural compounds (NCs) with PA *vs*. AD and PD. Considering the role of neuroinflammation in the onset and development of neurodegenerative pathologies, NCs with anti-inflammatory (AI) activity (AIA) could be good candidates for developing effective therapeutic strategies. Marine organisms represent a huge source of NCs and, among them, algae are appreciated sources of important bioactive NCs, for example, antioxidants (AOs), proteins, vitamins (Vits), minerals, soluble dietary fibers (SDFs), polyunsaturated fatty acids (PUFAs), polysaccharides (PSs), sterols, carotenoids, tocopherols, terpenes, phycobilins, phycocolloids, and phycocyanins. Numerous AI NCs were isolated from marine algae with potential protective efficacy *vs*. neuroinflammation.

Phycotoxins chemistry and biochemistry were reviewed [1]. It was informed total phenolic NCs, radical scavenging, and metal chelation of extracts from Icelandic seaweeds [2]. High correlation was found between total phenolic content (TPC) of seaweed extracts, and their scavenging capacity *vs*. 2,2-diphenyl-1-picrylhydrazyl (DPPH) and hydroperoxyl radicals $HOO^•$, indicating an important role of algal polyphenols as chain-breaking AOs. However, H_2O extracts had higher Fe^{2+}-chelating activity than 70% acetone CH_3COCH_3 extracts, and no correlation was found with their TPC, suggesting that other components, for example, PSs, proteins, and peptides, in the extracts were more effective chelators of Fe^{2+} than phenolic NCs. Marine-algae AIAs in neurodegenerative diseases were reviewed [3]. It was informed of the chemical diversity of headspace and volatile oil composition of two BAs (*Taonia atomaria* and *Padina pavonica*) from the Adriatic Sea [4]. It was reported AO activity (AOA), TPC, and total flavonoid content (TFC) of selected Kenyan medicinal plants, sea algae, and medicinal wild mushrooms [5]. Biochemical, micronutrient, and physicochemical properties of the dried red seaweeds *Gracilaria edulis* and *G. corticata* were published [6]. Potential AI NCs from marine algae was revised [7].

Earlier publications in *Nereis*, etc. classified yams [8], lactic acid bacteria [9], fruits [10], food spices [11], and legumes [12–14] by principal component, cluster, and meta-analyses. It was informed the molecular classifications of 33 phenolics NCs derived from the cinnamic and benzoic acids from *Posidonia oceanica* [15], 74 flavonoids [16], 66 stilbenoids [17], 71 triterpenoids and steroids from *Ganoderma* [18] and 17 isoflavonoids from *Dalbergia parviflora* [19], artemisinin derivatives [20, 21]. A tool for interrogation of the macromolecular structure was published [22]. Mucoadhesive polymer hyaluronan favors transdermal penetration absorption of caffeine [23, 24]. Endocrine disruptor diethylstilbestrol (DES), bisphenol-A (similar to DES), polycarbonate, and epoxy-silica nanocomposites were published [25, 26]. It was informed of the molecular classification of algae by phenolics and AOs [27]. The purpose of the present report is to review phycotoxin, commercial algal products, uses, phycoremediation, OS-PA, and AMA of loliolide and Pj-EE, and fucoxanthin-rich BAs extract improving male reproductive function on SZT-NAM-induced diabetic rat model.

7.2 PHYCOTOXIN, COMMERCIAL ALGAL PRODUCT, USES, AND PHYCOREMEDIATION

A few algae play a *nuisance role* where they tend to block the water supply canals filtration unit because of eutrophication of reservoirs, canals, or other water units [28]. Sometimes, the algae are labeled as *phycotoxin* as they secrete some toxins, which cause poisoning.

Commercial algal products and uses. Spirulina (arthrospira): human and animal nutrition, cosmetics (phycobiliproteins, powders, extracts, tablets, beverages, chips, pasta, liquids extract). *Chlorella* sp.: human nutrition, aquaculture, cosmetics (tablets, powders, nectar, noodles). *Dunaliella salina*: human nutrition, cosmetics (β-carotene, powders). *Aphanizomenon flos-aquae*: human nutrition (capsules, crystals, powder). *Haematococcus pluvialis*: aquaculture, astaxanthin. *Crypthecodinium cohnii* and *Shizochytrium* sp. docosahexaenoic acid (DHA) oil.

The use of micro or macroalgae to remove pollutants and nutrients from the wastewater is called phycoremediation.

7.3 OS-PROTECTIVE/ANTI MELANOGENIC EFFECTS OF LOLIOLIDE FROM *P. JAPONICA*

The GAs are considered as a representative biosource to be applied for the preparation of pharmaceutical, nutraceutical, and cosmeceutical products. *Prasiola japonica* is freshwater GAs with a lot of NCs, for example, MTL γ-lactone (-)-loliolide (*cf.* Figure 7.1), methylpyrazine, 1-hydroxy-2-propanone, diisopropylamine, 1,6-dihydro-6-oxo-3-pyridinecarboxamide, mannitol, mannose, and glucitol, according to gas chromatography/mass spectrometric (GC/MS) analysis. Of them, loliolide results in an active component in GAs with a number of bioroles (e.g., anti-aging, antiviral, AIAs). However, the effects of freshwater GAs components, loliolide, and Pj-EE, on the skin were not studied extensively. The OS-protective effects of loliolide and Pj-EE were investigated in human keratinocyte HaCaT cells [29]. To do this, the effects of loliolide and Pj-EE on the expression of AO protein nuclear factor (erythroid-derived 2)-like 2 (NRF2)-Kelch-like ECH-associated protein 1 (KEAP1) signaling pathway proteins were confirmed by Western blot. The expression of genes encoding matrix metalloproteinases (MMPs) was confirmed by reverse transcription-polymerase chain reaction (RT-PCR) and real-time PCR analyses, and cell viability was determined by 3-(4,5-dimethylthiazol-2-yl)-2,5-diphenyltetrazolium bromide (MTT) assay. Mouse melanoma B16F10 cells were used to assess AMA of loliolide and Pj-EE, measuring the level of melanin content and secretion. The expression of the melanocortin-1 receptor (MC1R) proteins related to melanogenesis was confirmed by Western blot.

7.4 FUCOXANTHIN-RICH BROWN ALGAE IMPROVE MALE REPRODUCTIVE FUNCTION

Hypogonadism and OS are occurring commonly in men with diabetes and associated male infertility. The capability of AOA and AIA of fucoxanthin were investigated, and its protective effects on male reproduction in diabetic rats were evaluated [30]. The RAW 264.7 macrophage cells were used to evaluate AO and AI activity. Thirty male Sprague–Dawley rats were induced by SZT–NAP for a diabetes model and fed either with three different doses of fucoxanthin or rosiglitazone for four weeks. The fucoxanthin significantly inhibited NO production and reduced reactive oxygen species (ROSs) level in lipo-PS (LPS)-induced RAW 264.7 cells. In the animal study, fucoxanthin administration improved insulin resistance, restored sperm motility, decreased abnormal sperm number, and inhibited lipid peroxidation. It restored G-protein coupled receptor (GPR)54 (KISS1 receptor) and suppressor of cytokine signaling (SOCS)3 messenger ribonucleic acid (mRNA) expression in the hypothalamus and recovered luteinizing hormone level and the testosterone level. Fucoxanthin not only possessed AO and AI properties but also decreased the diabetes signs and symptoms, and improved spermatogenesis and male reproductive function.

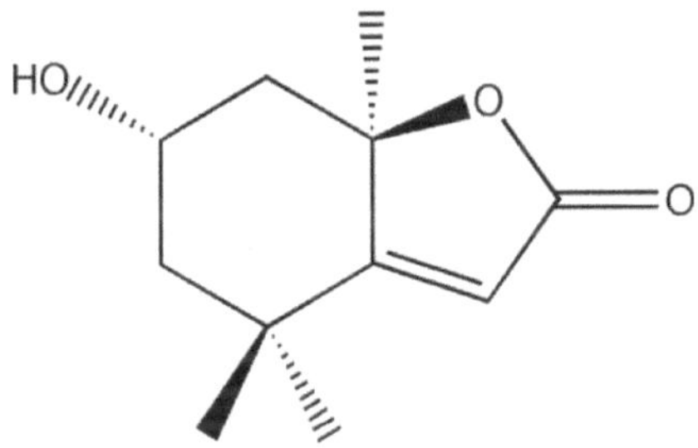

FIGURE 7.1 Molecular structure of loliolide.

7.5 DISCUSSION

Neuroinflammation is the main contributor to the onset and progression of neurodegenerative diseases, for example, AD and PD. Microglial and astrocyte activation is a brain defense mechanism to counteract harmful pathogens and damaged tissues, while their prolonged activation induces neuroinflammation that triggers or exacerbates neurodegeneration. No pharmacological therapies exist able to slow down or stop the progression of neurodegeneration. Research is turning to the identification of NCs with PA *vs*. the diseases. Considering the role of neuroinflammation in the onset and development of neurodegenerative pathologies, NCs with AIA could be good candidates for developing effective therapeutic strategies. Marine organisms represent a huge source of NCs and, among them, algae are appreciated sources of important bioactive components, for example, AOs, proteins, Vits, minerals, SDFs, PUFAs, PSs, sterols, carotenoids, tocopherols, terpenes, phycobilins, phycocolloids, and phycocyanins. Numerous AI NCs were isolated from marine algae with potential protective efficacy *vs.* neuroinflammation.

7.6 FINAL REMARKS

From the present results and discussion, the following final remarks can be drawn.

1. Challenges of algae follow: (a) improvement in the systems of search and detection of algae species and active compounds; (b) improvement in the knowledge of algae genome to know production routes of different compounds; (c) rational use of information with the objective of a sustainable production of foods, health, and energy; (d) optimization of the processes of production of biomass (*via* the design of efficient bioreactors); (e) process optimization of algae-compounds recovery/purification (*via* environmentally friendly processes/solvents); (f) design of biorefineries that allow the rational use of all produced resources, for example, residues reuse; (g) search for applications in which algae (or residues) be economically profitable beside tradition; (h) design of economic/ecological strategies that allow improving the joint work of different species.
2. The wide range of biological and bioactive molecules found in marine algae represents a challenge for researchers involved in the study of neuroinflammation and neurodegeneration processes. Marine algae extracts and many constituents belonging to different chemical classes were shown to exert preventive/protective effects *vs.* neuroinflammation; for example, they were shown to be effective in reducing inflammatory mediators (e.g., nitric oxide, tumor necrosis factor-α, interleukin-6, interleukin-1β), in downregulating inflammatory enzymes (e.g., inducible nitric oxide synthase, *cyclo*-oxygenase-2) and in modulating the signaling pathways that lead to the activation of nuclear factor κ-light-chain-enhancer of activated B cells. Most of the compounds isolated from marine algae showed antioxidant activity. Oxidative stress represents a hallmark of neuroinflammation and its counteraction was a strategy in the prevention of neurodegeneration.

3. Fucoxanthin not only possessed antioxidant and anti-inflammatory properties but also decreased the diabetes signs and symptoms, and improved spermatogenesis and male reproductive function.

ACKNOWLEDGMENTS

The authors thanks the support from Fundacion Universidad Catolica de Valencia San Vicente Martir (Project No. 2019-217-001UCV).

KEYWORDS

- **anti-melanogenesis**
- **antioxidant**
- **antioxidant activity**
- **icelandic seaweed**
- **loliolide**
- ***Prasiola japonica***
- **total phenolic content**

REFERENCES

1. Botana, L. M., (2007). *Phycotoxins: Chemistry and Biochemistry*. Blackwell: Ames, IA.
2. Wang, T., Jónsdóttir, R., & Ólafsdóttir, G., (2009). Total phenolic compounds, radical scavenging and metal chelation of extracts from Icelandic seaweeds. *Food Chem., 116*, 240–248.
3. Barbalace, M. C., Malaguti, M., Giusti, L., Lucacchini, A., Hrelia, S., & Angeloni, C., (2019). Anti-inflammatory activities of marine algae in neurodegenerative diseases. *Int. J. Mol. Sci., 20,* 3061–1–20.
4. Jerkovic, I., Kranjac, M., Marijanovic, Z., Roje, M., & Jokic, S., (2019). Chemical diversity of headspace and volatile oil composition of two brown algae (*Taonia atomaria* and *Padina pavonica*) from the Adriatic Sea. *Molecules, 24*, 495–1–12.
5. Siangu, B. N., Sauda, S., John, M. K., & Njue, W. M., (2019). Antioxidant activity, total phenolic and flavonoid content of selected Kenyan medicinal plants, sea algae and medicinal wild mushrooms. *Afr. J. Pure Appl. Chem., 13*, 43–48.
6. Rosemary, T., Arulkumar, A., Paramasivam, S., Mondragon-Portocarrero, A., & Miranda, J. M., (2019). Biochemical, micronutrient and physicochemical properties of the dried red seaweeds *Gracilaria edulis* and *Gracilaria corticata*. *Molecules, 24*, 2225–1–14.

7. Fernando, I. P. S., Nah, J. W., & Jeon, Y. J., (2016). Potential anti-inflammatory natural products from marine algae. *Environ. Toxicol. Pharmacol., 48*, 22–30.
8. Torrens-Zaragozá, F., (2013). Molecular categorization of yams by principal component and cluster analyses. *Nereis, 2013*(5), 41–51.
9. Torrens-Zaragozá, F., (2014). Classification of lactic acid bacteria against cytokine immune modulation. *Nereis, 2014*(6), 27–37.
10. Torrens-Zaragozá, F., (2015). Classification of fruits proximate and mineral content: Principal component, cluster, meta-analyses. *Nereis, 2015*(7), 39–50.
11. Torrens-Zaragozá, F., (2016). Classification of food spices by proximate content: Principal component, cluster, meta-analyses. *Nereis, 2016*(8), 23–33.
12. Torrens, F., & Castellano, G., (2014). From Asia to the Mediterranean: Soya bean, Spanish legumes, and commercial *soya bean* principal component, cluster, and meta-analyses. *J. Nutr. Food Sci., 4*(5), 98–98.
13. Torrens, F., & Castellano, G., (2018). Principal component, cluster, and meta-analyses of soya bean, Spanish legumes, and commercial soya bean. In: Torrens, F., Balköse, D., & Thomas, S., (eds.), *High-Performance Materials and Engineered Chemistry* (pp. 267–294). Apple Academic–CRC: Waretown, NJ.
14. Torrens, F., & Castellano, G., (2019). Meta-analysis of underutilized beans with nutrients profile. In: Aguilar, C. N., & Carvajal-Millan, E., (eds.), *Applied Food Science and Engineering with Industrial Applications* (pp. 247–277). Apple Academic–CRC: Waretown, NJ.
15. Castellano, G., Tena, J., & Torrens, F., (2012). Classification of polyphenolic compounds by chemical structural indicators and its relation to antioxidant properties of *Posidonia oceanica* (L.) delile. *MATCH Commun. Math. Comput. Chem., 67*, 231–250.
16. Castellano, G., González-Santander, J. L., Lara, A., & Torrens, F., (2013). Classification of flavonoid compounds by using entropy of information theory. *Phytochemistry, 93*, 182–191.
17. Castellano, G., Lara, A., & Torrens, F., (2014). Classification of stilbenoid compounds by entropy of artificial intelligence. *Phytochemistry, 97*, 62–69.
18. Castellano, G., & Torrens, F., (2015). Information entropy-based classification of triterpenoids and steroids from *Ganoderma*. *Phytochemistry, 116*, 305–313.
19. Castellano, G., & Torrens, F., (2015). Quantitative structure–antioxidant activity models of isoflavonoids: A theoretical study. *Int. J. Mol. Sci., 16*, 12891–12906.
20. Torrens, F., Redondo, L., & Castellano, G., (2017). Artemisinin: Tentative mechanism of action and resistance. *Pharmaceuticals, 10*, 20–44.
21. Torrens, F., Redondo, L., & Castellano, G., (2018). Reflections on artemisinin, proposed molecular mechanism of bioactivity and resistance. In: Haghi, A. K., Balköse, D., & Thomas, S., (eds.), *Applied Physical Chemistry with Multidisciplinary Approaches* (pp. 189–215). Apple Academic–CRC: Waretown, NJ.
22. Torrens, F., & Castellano, G., (2014). A tool for interrogation of macromolecular structure. *J. Mater. Sci. Eng. B, 4*(2), 55–63.
23. Torrens, F., & Castellano, G., (2014). Mucoadhesive polymer hyaluronan as biodegradable cationic/zwitterionic–drug delivery vehicle. *ADMET DMPK, 2*, 235–247.
24. Torrens, F., & Castellano, G., (2015). Computational study of nanosized drug delivery from cyclodextrins, crown ethers and hyaluronan in pharmaceutical formulations. *Curr. Top. Med. Chem., 15*, 1901–1913.

25. Torrens-Zaragozá, F., (2011). Polymer bisphenol-A, the incorporation of silica nanospheres into epoxy-amine materials and polymer nanocomposites. *Nereis, 2011*(3), 17–23.
26. Torrens, F., & Castellano, G., (2013). Bisphenol, diethylstilbestrol, polycarbonate and the thermomechanical properties of epoxy–silica nanostructured composites. *J. Res. Upd. Polym. Sci., 2*, 183–193.
27. Torrens, F., & Castellano, G., (2019). Molecular classification of algae by phenolics and antioxidants. In: Aguilar, C. N., & Carvajal-Millan, E., (eds.), *Applied Food Science and Engineering with Industrial Applications* (pp. 279–303). Apple Academic–CRC: Waretown, NJ.
28. Singh, R., Joshi, H., & Singh, A., (2020). Algae: Role in environment safety. In: Pourhashemi, A., Deka, S. C., & Haghi, A. K., (eds.), *Research Methods and Applications in Chemical and Biological Engineering* (pp. 237–248). Apple Academic–CRC: Waretown, NJ.
29. Park, S. H., Choi, E., Kim, S., Kim, D. S., Kim, J. H., Chang, S., Choi, J. S., et al., (2018). Oxidative stress-protective and anti-melanogenic effects of loliolide and ethanol extract from freshwater green algae, *Prasiola japonica. Int. J. Mol. Sci., 19*, 2825–1–15.
30. Kong, Z. L., Sudirman, S., Hsu, Y. C., Su, C. Y., & Kuo, H. P., (2019). Fucoxanthin-rich brown algae extract improves male reproductive function on streptozotocin–nicotinamide-induced diabetic rat model. *Int. J. Mol. Sci., 20*, 4485–1–17.

CHAPTER 8

Recent Advancements in the Pharmacotherapeutic Perspectives of Some Chalcone Scaffold Containing Natural Compounds as Potential Anti-Virals

DEBARSHI KAR MAHAPATRA,[1] SANJAY KUMAR BHARTI,[2] and VIVEK ASATI[3]

[1]*Department of Pharmaceutical Chemistry, Dadasaheb Balpande College of Pharmacy, Nagpur, Maharashtra, 440037, India,*
E-mail: mahapatradebarshi@gmail.com

[2]*Institute of Pharmaceutical Sciences, Guru Ghasidas Vishwavidyalaya (A Central University), Bilaspur, Chhattisgarh, 495009, India*

[3]*Department of Pharmaceutical Chemistry, NRI Institute of Pharmacy, Bhopal, Madhya Pradesh, 462021, India*

ABSTRACT

In spite of the development of significant elementary acquaintance regarding a variety of pathogens, the existing chemotherapy is still unacceptable as a result of imperfect effectiveness, long-term action, high cost, and unwanted side-effects. Natural products have been found to express numerous pharmacological activities. Several alcoholic, hydroalcoholic, and aqueous extracts along with phytoconstituents are reported to exhibit anti-infective effects by modulating the number of pharmacotherapeutic targets. The present chapter focuses on the therapeutic perspectives of chalcone based natural compounds in exhibiting antiviral activity against rhinovirus, human immunodeficiency virus (HIV), influenza virus, poliovirus, dengue virus, and tobacco mosaic virus (TMV) by modulating vital therapeutic targets like HIV-1 protease, DEN-1 NS3 protease, and neuraminidase. These natural compounds serve as interesting leads for synthetic drug fabrication by various drug design, drug discovery, and drug development approaches as well as structure-activity relationships (SARs) predictions. These compounds from nature may serve as templates for further exploration of molecular targets, structural biology, mechanism of actions (MOAs), etc.

8.1 INTRODUCTION

In the latest years, extensive outburst of abundant infectious ailments across the world has created chaos between the populations [1]. Chiefly, the residents in the tropical and sub-tropical areas are concerned by these virulent pathogens [2]. In several nations diseases of protozoan origin (sleeping sickness, leishmaniasis, amoebiasis, etc.), *Plasmodium* origin (*P. knowlesi*, *P. ovale*, *P. vivax*, *P. malariae*, *P. falciparum*, etc.), microbial origin (tuberculosis, dermatophyte infection, cholera, etc.), viral origin (HIV, Ebola Virus Disease, Dengue, AIDS, Rhinovirus, Influenza, Rotavirus, Mosquito-borne viruses, poliovirus, etc.), worms origin, etc. have been perceived by the world which causes drastic inflammation in liver, normal muscle and joints damage, acute hair loss, reduced hearing ability, etc. [3]. The products of employing individual compounds stay challenging as a result of resistance to the parent molecule [4]. A great bacterial population dwells in the body due to an appropriate atmosphere that offers a slighter impediment for development, a huge region for connection, and a superior foundation of sustenance [5]. The attained virulence and encompassed host defense gets frequently aggravated to the "opportunistic" pathogens, which in the due way of time damages the offered performance of the human body [6]. The microscopic pathogens stay over humid, damp, moist, and dark regions of the human body such as hairs, skin, dead tissues, and nails which affects the local population of all sexes and ages, worldwide [7].

In spite of the development of significant elementary acquaintance regarding a variety of pathogens, the existing chemotherapy is still unacceptable as a result of imperfect effectiveness, long-term action, high cost, and unwanted side-effects [8]. The rise of drug resistance by these virulent pathogens, especially in the viral-infected patients has also forced the health difficulties [9]. Thus, there is a critical need for the upgrading of new, competent, and harmless drugs against these organisms. Natural products and traditional medicines are boon for the complete management or treatment of several opportunistic infections [10].

Natural products have been found to express numerous pharmacological activities. Several alcoholic, hydroalcoholic, and aqueous extracts along with phytoconstituents are reported to exhibit anti-infective effects by modulating the number of pharmacotherapeutic targets [11]. The present chapter focuses on the therapeutic perspectives of chalcone based natural compounds in exhibiting antiviral activity against rhinovirus, human immunodeficiency virus (HIV), influenza virus, poliovirus, dengue virus, and tobacco mosaic virus (TMV) by modulating vital therapeutic targets like HIV-1 protease, DEN-1 NS3 protease, and neuraminidase.

8.2 CHALCONES

Chalcones (a precursor of isoflavonoid and flavonoid) or 1,3-diphenyl-2-propene-1-one (an open-chain intermediate in the aurone synthesis) or benzylideneacetophenone (comprising of two aromatic components linked by an α, β-unsaturated carbonyl bridge function) is one of the most privileged scaffolds (due to their simplicity in the overall product chemistry, the feasibility of productive computational studies, ten replaceable hydrogen atoms for producing multiple active compounds, and therapeutic targets modulating ability) originated from Mother Nature that can be successfully synthesized in laboratory scale (first fabricated as chromophoric products by Kostanecki and Tambor in 19th century from benzaldehyde and acetophenone) [12, 13]. A number of recently reported and well-established traditional high-yield methods such as Claisen-Schmidt reaction, Suzuki-Miyaura reaction, Friedel-Crafts reaction, Julia-Kocienski reaction, One-pot reactions, Sonogashira isomerization coupling, Microwave-assisted reactions, Direct crossed-coupling reaction, Carbonylated Heck coupling reaction, Solvent-free reactions, Solid acid catalyst mediated reactions, etc. have been into academic-scale and industrial-scale applications [14, 15].

These natural low molecular weight ligands convert into flavonoid (by acid) and reverse back into flavanone (by base) and find application in the structural elucidation of chromanochromane, tannin, flavanone, and flavonoid. In addition to it, they are an excellent template for synthesizing numerous pharmacotherapeutically active synthetic scaffolds like benzodiazepine, thiadiazole, benzoxazepine, benzothiazepine, pyrazole, isoxazole, pyrimidine, pyrazoline, etc., which expressed multifarious actions against protozoal infections, bacterial infections, sleep disorder, trypanosomiasis, anxiety, gout, malaria, epilepsy, hypertension, reduced immune response, diabetes, tuberculosis, leishmaniasis, ulcer, fungal infections, cancer, thrombosis, reactive oxygen species, inflammation, HIV, metastasis, etc. [16, 17]. Non-pharmacological applications in daily human life rest with a scintillator, insecticides, polymerization agents in product development, sweeteners in confectionaries, catalyst in specific reactions, chemosensor for detection, fluorescent polymeric agents, chromophores in dying industries, fluorescent whitening agent, etc. [18].

8.3 CHALCONES AS ANTI-HIV AGENTS

Human immunodeficiency virus (HIV) is a challenge for decades and has affected 40 million people till date. Millions of people lost their lives as the virus targeted the immune system of humans and due to the weakened defense systems against the approaching infections, numerous cancer forms, and other stern clinical expressions (CD cell count). The major global public health issue impaired the normal functioning of the immune cells and the immunodeficient individual is classified as acquired immunodeficiency syndrome (AIDS) which takes nearly 15 years to develop. Africa and sub-Saharan Africa are the chiefly affected constituency with more than 30 million (60% global infections) HIV affected people individuals [19]. A schematic diagram is provided for a plausible mechanism of chalcone action on HIV targets in Figure 8.1.

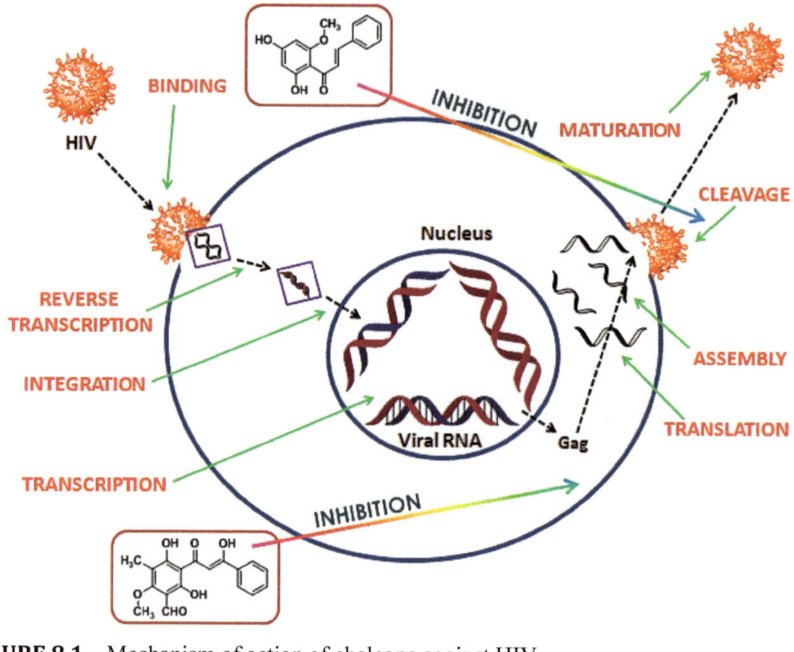

FIGURE 8.1 Mechanism of action of chalcone against HIV.

8.3.1 *HIV REPLICATION INHIBITORS*

2-methoxy-3-methyl-4,6-dihydroxy-5-(3'-hydroxy)-cinnamoylbenzaldehyde **(2)** was a natural product isolated from *Desmos species* have been found to potentially inhibit the HIV replication in H9 lymphocyte cells (Figure 8.2) [20].

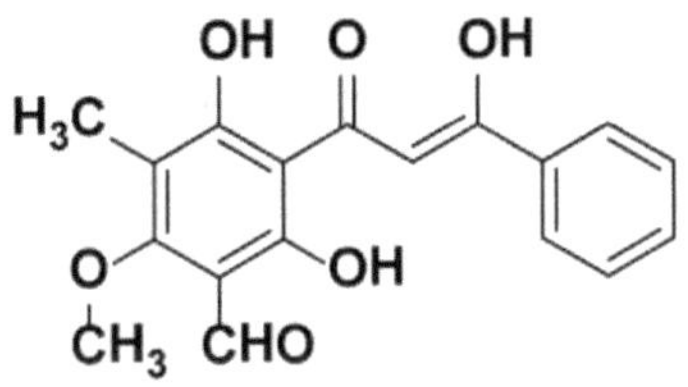

(2)

FIGURE 8.2 Structure of HIV-replication inhibitor.

8.3.2 *HIV-PROTEASE INHIBITORS*

HIV protease is an enzyme of proteolytic nature that plays an imperative role in the formation of structural proteins and enzymes. This homodimer essentially breaks the huge viral protein components that have numerous functions in substrate binding, terminal replication steps, cellular progeny maturation, etc. There are two types of proteases that play functional dominance in the biochemical process. The first class of protease enzymes employs an activated water molecule to attack the amide bond carbonyl of the substrate's scissile bond. In the second class of proteases, a nucleophilic atom of an amino acid side chain is employed to commence amide hydrolysis. Chalcones have been seen to reversibly bind with the protease active site and prevent the proteolytic cleavage steps that metamorphosize the inactive virus components into their fully mature infectious forms. Treatment with protease inhibitors has resulted in 1000 times decrease in plasma HIV levels within 3 months duration [21].

From *Boesenbergia pandurata* Holtt, cardamonin **(3)** was isolated and screened where the chalcone molecule successfully inhibited the HIV-1 protease (HIV-PR) with IC_{50} value of 31 mg/mL (Figure 8.3) [22].

8.4 CHALCONES AS ANTI-POLIO VIRUS AGENTS

The extremely infectious disease invading the nervous system, known as "Polio," is primarily responsible for total paralysis (often irreversible in

1 among 200 people) of several body parts (5–10% individuals die when the breathing muscles become immobilized) in a very short time. It chiefly affects children below the age of 5 yrs in developing nations (predominantly; Afghanistan, Pakistan, and Nigeria) and spreads essentially from person to person through contaminated food, fecal-oral route, etc. in the intestines [23]. The mechanism of poliovirus growth and the potentials of chalcones in expressing anti-polio virus actions are depicted in Figure 8.4.

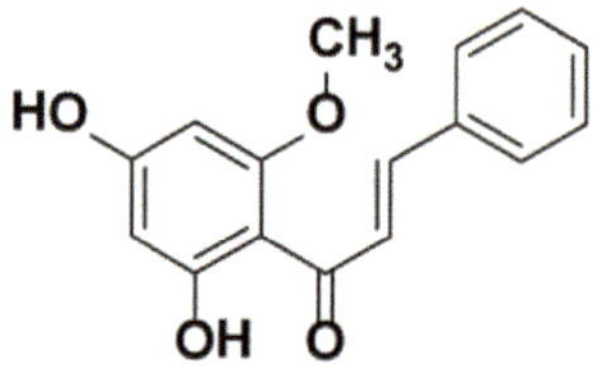

(3)

FIGURE 8.3 Structure of HIV-Protease inhibitor.

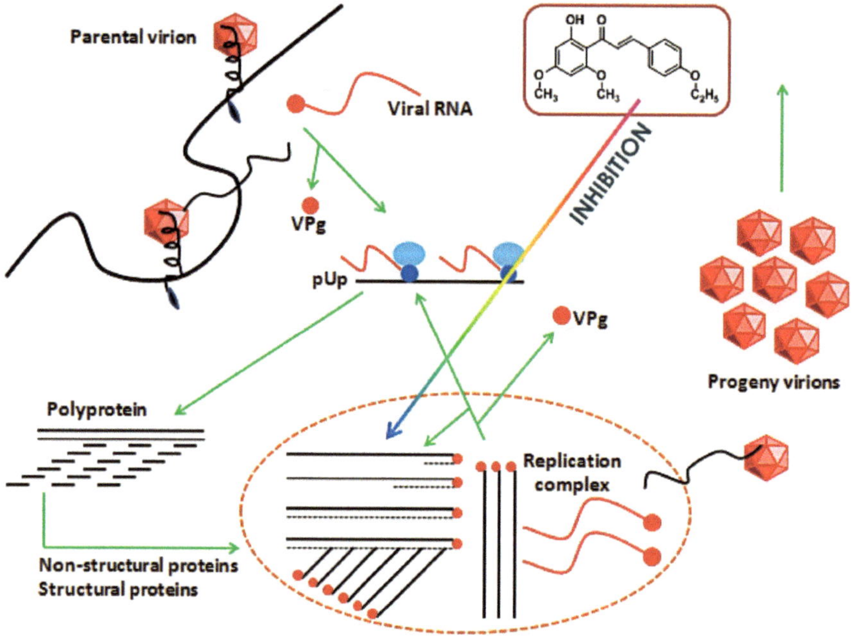

FIGURE 8.4 Mechanism of action of chalcone against poliovirus.

The anti-polio viral activity was explored by a group of active researchers for a plant-derived chalcone product, 4'-ethoxy-2-hydroxy-4,6-dimethoxy chalcone (4) against two imperative cell lines RD and L2OB where at the concentration 0.03 mg/mL demonstrated remarkable inhibition of the cell growth by modulating the therapeutic target(s) (Figure 8.5) [24].

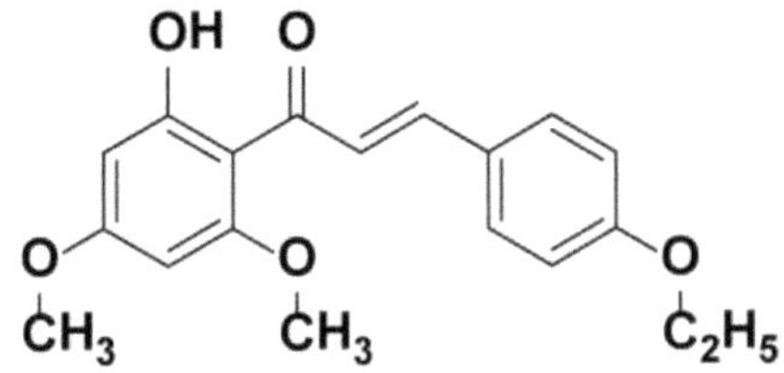

(4)

FIGURE 8.5 Structure of chalcone-based anti-polio virus agent.

8.5 CHALCONES AS ANTI-RHINO VIRUS AGENTS

Common cold or seasonal influenza is a contagious global disease of viral origin (rhinovirus) that affects every individual. Based on the virus's surface proteins, it is classified into three types: Influenza A virus (high frequency), Influenza B virus, and Influenza C virus (least frequency). Influenza A virus is further sub-classified into two sub-types: hemagglutinin (H) and neuraminidase (N); and further they are classified into numerous strains. The subtypes H1N1 and H3N2 have been perceived to be the most currently active form that had already spread pandemic in several developing nations in the last few years during the winters. Overall, the virus is a serious public issue and causes death to millions of people [25]. Chalcones are known to inhibit the proliferation of the virus. A schematic diagram is provided for a plausible mechanism of chalcone action in expressing anti-polio virus activity in Figure 8.6.

Chalcone molecule, 4-ethoxy-2-hydroxy-4',6-dimethoxychalcone (Ro 09-0410) (5) has been recognized as anti-picornavirus and anti-rhinovirus agent. In HeLa cell cultures, out of the tested 53 rhinovirus serotypes, the chalcone Ro 09-0410 presented susceptibility against 46 serotypes in a temperature-dependent, time-dependent, and dose-dependent manner. The novel chalcone inhibited half of the rhinovirus serotypes at 0.03 mg/mL whereas expressing about 50% cytotoxicity at a concentration of 30 mg/mL (Figure 8.7) [26].

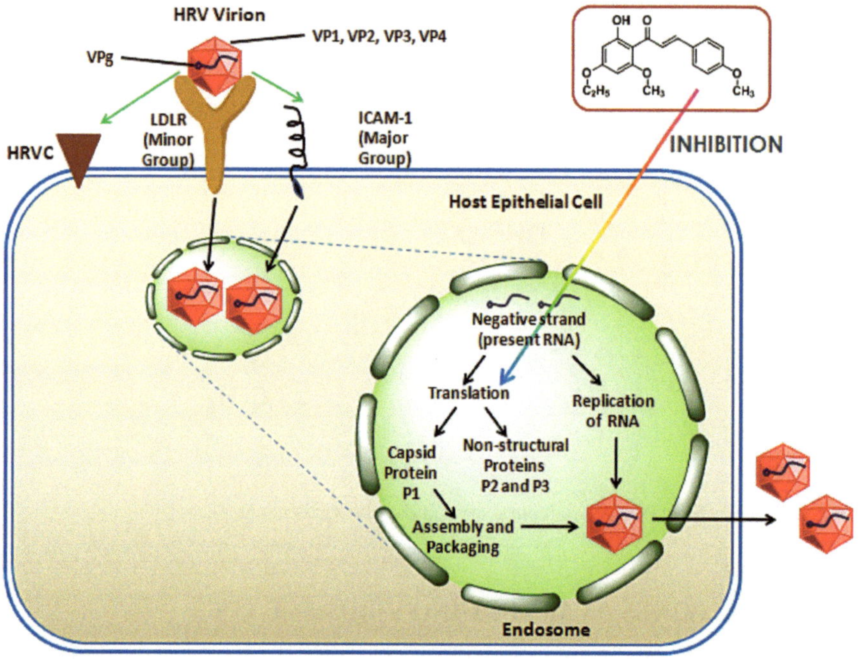

FIGURE 8.6 Mechanism of action of chalcone against rhinovirus.

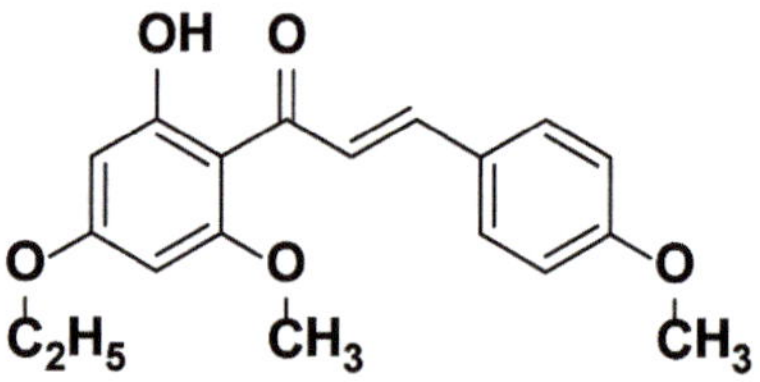

(5)

FIGURE 8.7 Structure of chalcone-based anti-rhino virus agent.

8.6 CHALCONES AS ANTI-DENGUE VIRUS AGENTS

8.6.1 DEN-1 NS3 PROTEASE INHIBITORS

The well-known RNA helicase, serine protease, RTPase, and NTPase; DENV NS3 comprises of the catalytic region having His-51, Asp-75, and Ser-135 amino acid component between the β-strands which plays a dominant role in the proteolytic processing of the structural proteins by signaling peptides situated in the endoplasmic reticulum. The catalytic triad is active only in the presence of the NS2B cofactor that wraps the protease domain. The DENV helicase sub-domains (I, II, and III) are composed of NS3 residues (180-618). The dengue virus type 2 (DEN 2) containing 10.7 kb single-stranded RNA genome of order NH2-C-prM-E-NS1-NS2A-NS2B-NS3-NS4A-NS4B-NS5-COOH has an imperative role in the proliferation of dengue virus. Inhibition of this target leads to the cessation of dengue viral growth. A schematic diagram is provided for a plausible mechanism of chalcone action on DEN-1 NS3 inhibition in Figure 8.8 [27].

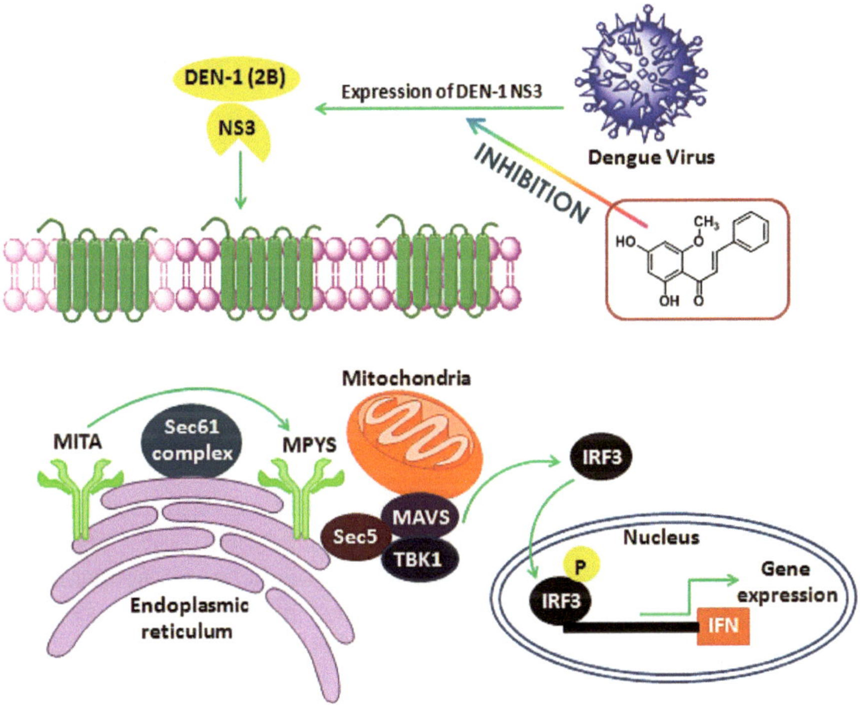

FIGURE 8.8 Mechanism of action of chalcone against dengue virus.

Cardamonin (3), a well-known phytochalcone isolated from *Boesenbergia rotunda* expressed rapid inhibition of Dengue Type-2 virus growth by inhibiting the NS3 protease in a non-competitive way and also demonstrated synergistic action in growth inhibitory function when administered along with pinocembrin (Figure 8.9) [28].

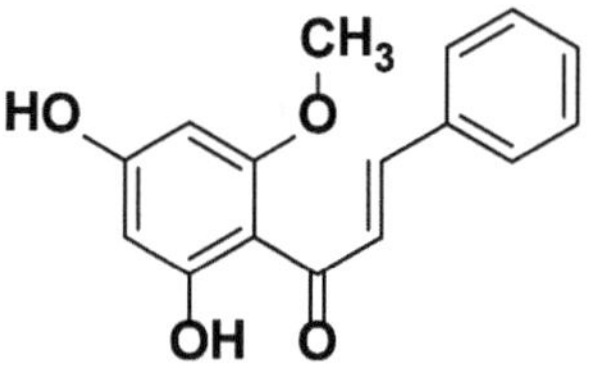

(3)

FIGURE 8.9 Structure of DEN-1 NS3 protease inhibitor.

8.7 CHALCONES AS ANTI-INFLUENZA VIRUS AGENTS

8.7.1 NEURAMINIDASE INHIBITORS

The enzyme class neuraminidases are situated on the influenza virus surface which is primarily responsible for the release of developed matured viruses from the host cells by selectively cleaving the sialic acid components present in glycoproteins (located on influenza virus). The sialic acid is responsible for the active viral replication and binding with the interior part of host cell surfaces (through hemagglutinin). Neuraminidases cleaved the sialic acid groups from the host cell and the influenza virus replication initiates. Inhibition of the neuraminidase enzyme by the chalcone compounds is an effective way to manage influenza symptoms. A schematic diagram is provided for a plausible mechanism of chalcone action on neuraminidase inhibition in Figure 8.10 [29].

From the buds of *Cleistocalyx operculatus*, the isolated chalcone compounds presented a tremendous inhibition of both H1N1 and H9N2 viruses in the performed assay by selecting modulating the neuraminidase enzyme. The molecule (6–9) displayed the most impressive pharmacological action with IC_{50} values of 8.15 μM (against H1N1-WT strain) and 3.31 μM (H1N1-H274Y mutant strain) [30].

From the stem of *Glycyrrhiza inflate*, therapeutically active chalcones (10–11) were isolated and screened where prominent anti-neuraminidase activity was perceived by the investigators for non-prenylated candidates against the H1N1 strain, H9N2 strain, H1N1-WT strain, and H1N1-H274Y mutant strain in 293T cells [31].

Alkylated chalcones isolated from *Angelica keiskei* demonstrated the potential in inhibiting the influenza virus NA in a reversible non-competitive manner. The compound 2-hydroxy-3-methyl-3-butenyl alkyl (HMB) substituted chalcone (12) was found to be the most privileged inhibitor of neuraminidase (Figure 8.11) [32].

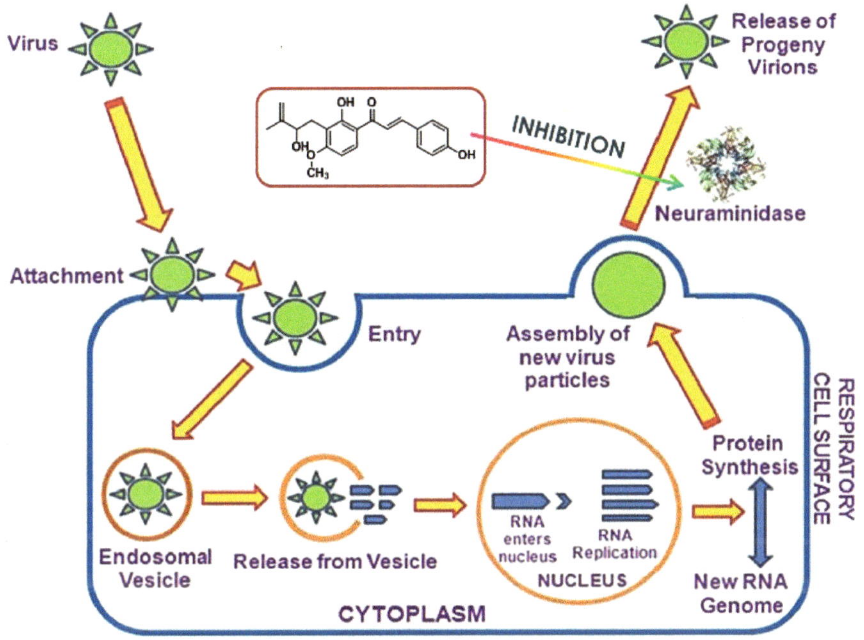

FIGURE 8.10 Mechanism of action of chalcone against influenza virus.

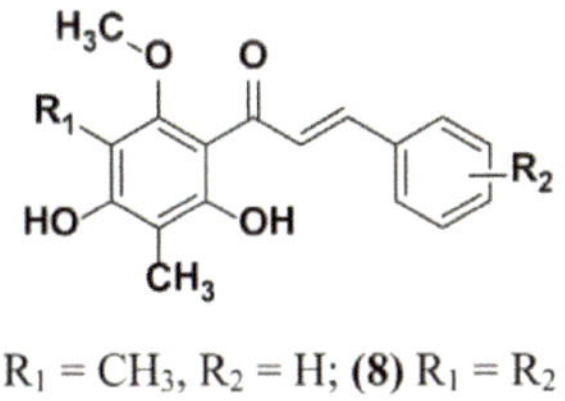

(6) $R_1 = CH_3$, R_2 = 4-OH; **(7)** $R_1 = CH_3$, $R_2 = H$; **(8)** $R_1 = R_2 = H$; **(9)** $R_1 = CH_3$, R_2 = 2-OH

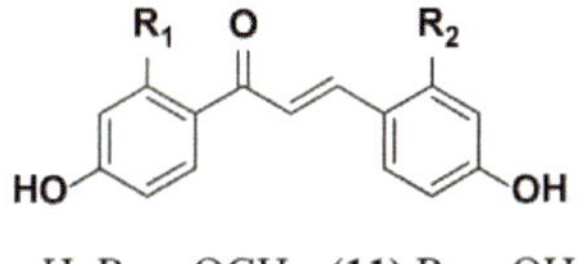

(10) $R_1 = H$, $R_2 = OCH_3$; **(11)** $R_1 = OH$, $R_2 = H$

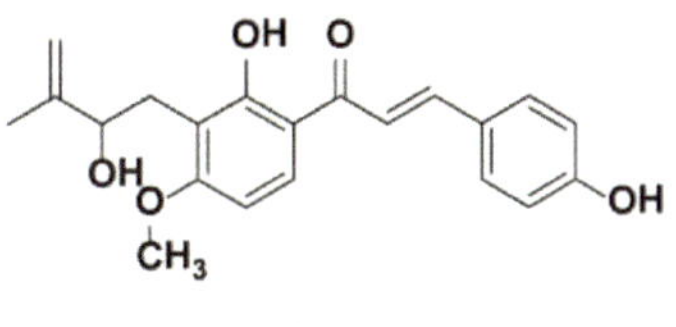

(12)

FIGURE 8.11 Structure of neuraminidase inhibitor.

8.8 CHALCONES AS ANTI-TMV AGENTS

This is a classic example that stated the role of chalcone compounds against the disease (Tobacco Mosaic Virus – TMV) which equally affects the plants (by causing stunting, mosaic, necrosis, mottling, yellowing of plant tissues, leaf curling, and ultimately death), rather than human. Chalcones have the potential as an anti-TMV agent by entirely wipe out the disease [33].

From *Desmodium podocarpum* whole plant, podocarchalcones (13–14) and other chalcones (15-16) have been isolated and screened for anti-tobacco mosaic virus activity (15.2–22.4% range) (Figure 8.12) [34].

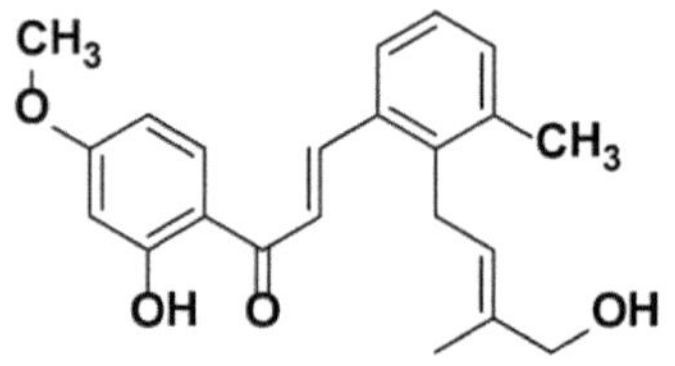

(13)

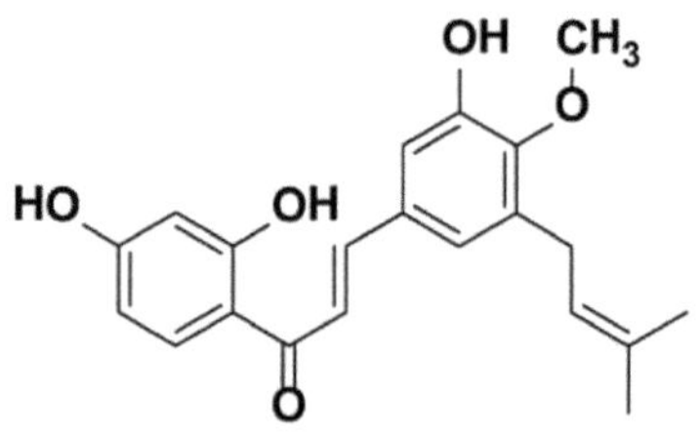

(14)

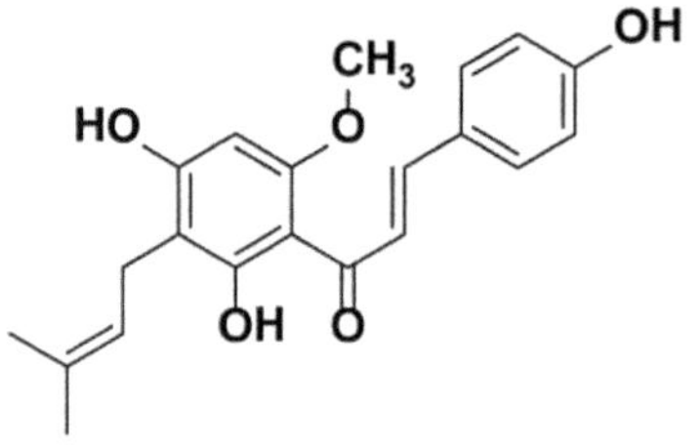

(15)

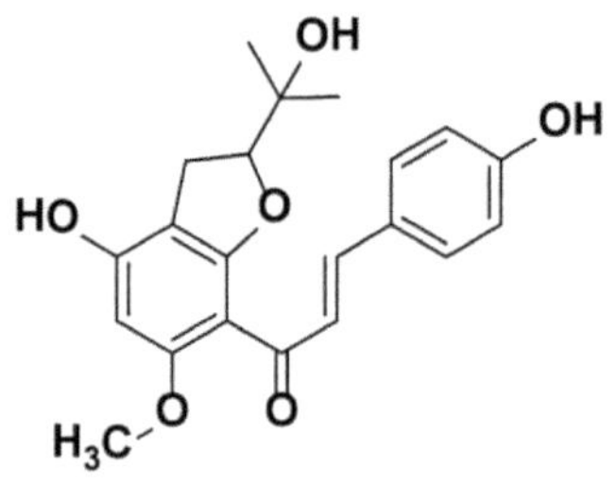

(16)

FIGURE 8.12 Structure of chalcone-based anti-TMV agent.

8.9 CONCLUSION

This comprehensively highlighted book chapter on the anti-viral (rhinovirus, poliovirus, retrovirus, SARS virus, dengue virus, and tobacco virus) perspectives of chalcone scaffold-based compounds by modulating diverse molecular targets such as DEN-1 NS3, protease, neuraminidase, etc. These natural compounds serve as interesting leads for synthetic drug fabrication by various drug design, drug discovery, and drug development approaches as well as structure-activity relationships (SARs) predictions. These compounds from nature may serve as templates for further exploration of molecular targets, structural biology, mechanism of actions (MOAs), etc.

KEYWORDS

- **anti-viral**
- **chalcone**
- **dengue**
- **HIV**
- **influenza**
- **polio**
- **rhino**
- **TMV**

REFERENCES

1. Mahapatra, D. K., & Bharti, S. K., (2016). *Drug Design*. New Delhi: Tara Publications Private Limited.
2. Shivhare, R. S., & Mahapatra, D. K., (2019). *Medicinal Chemistry-II*. Nagpur: ABD Publications Private Limited.
3. Mahapatra, D. K., & Bharti, S. K., (2017). *Handbook of Research on Medicinal Chemistry: Innovations and Methodologies*. New Jersey: Apple Academic Press.
4. Chhajed, S. S., Bastikar, V., Bastikar, A. V., & Mahapatra, D. K., (2019). *Computer Aided Drug Design*. Pune: Everest Publishing House.
5. Chhajed, S. S., Upasani, C. D., Wadher, S. J., & Mahapatra, D. K., (2017). *Medicinal Chemistry*. Nashik: Career Publications Private Limited.
6. Torrens, F., Mahapatra, D. K., & Haghi, A. K., (2019). *Biochemistry, Biophysics, and Molecular Chemistry: Applied Research and Interactions*. New Jersey: Apple Academic Press.
7. Puranik, M. P., & Mahapatra, D. K., (2019). *Medicinal Chemistry-III*. Nagpur: ABD Publications Private Limited.
8. Mahapatra, D. K., & Bharti, S. K., (2019). *Medicinal Chemistry with Pharmaceutical Product Development*. New Jersey: Apple Academic Press.

9. Mahapatra, D. K., & Bharti, S. K., (2020). *Pharmacological Perspectives of Low Molecular Weight Ligands*. New Jersey: Apple Academic Press.
10. Mahapatra, D. K., Bharti, S. K., & Asati, V., (2017). Nature inspired green fabrication technology for silver nanoparticles. *Curr. Nanomed., 7*(1), 5–24.
11. Mahapatra, D. K., Asati, V., & Bharti, S. K., (2019). Recent therapeutic progress of chalcone scaffold bearing compounds as prospective anti-gout candidates. *J. Crit. Rev., 6*(1), 1–5.
12. Mahapatra, D. K., Bharti, S. K., & Asati, V., (2015). Anti-cancer chalcones: Structural and molecular target perspectives. *European Journal of Medicinal Chemistry, 98*, 69–114.
13. Mahapatra, D. K., Bharti, S. K., & Asati, V., (2015). Chalcone scaffolds as anti-infective agents: Structural and molecular target perspectives. *European Journal of Medicinal Chemistry, 101*, 496–524.
14. Mahapatra, D. K., Asati, V., & Bharti, S. K., (2015). Chalcones and their therapeutic targets for the management of diabetes: Structural and pharmacological perspectives. *European Journal of Medicinal Chemistry, 92*, 839–865.
15. Mahapatra, D. K., & Bharti, S. K., (2016). Therapeutic potential of chalcones as cardiovascular agents. *Life Sciences, 148*, 154–172.
16. Mahapatra, D. K., Bharti, S. K., & Asati, V., (2017). Chalcone derivatives: Anti-inflammatory potential and molecular targets perspectives. *Current Topics in Medicinal Chemistry, 17*(28), 3146–3169.
17. Mahapatra, D. K., Bharti, S. K., Asati, V., & Singh, S. K., (2019). Perspectives of medicinally privileged chalcone based metal coordination compounds for biomedical applications. *European Journal of Medicinal Chemistry, 174*, 142–158.
18. Mahapatra, D. K., Asati, V., & Bharti, S. K., (2019). An updated patent review of therapeutic applications of chalcone derivatives (2014-present). *Expert Opinion on Therapeutic Patents, 29*(5), 385–406.
19. Moss, J. A., (2013). HIV/AIDS Review. *Radiologic Technology, 84*(3), 247–267.
20. Wu, J. H., Wang, X. H., Yi, Y. H., & Lee, K. H., (2003). Anti-AIDS agents 54. A potent anti-HIV chalcone and flavonoids from genus Desmos. *Bioorg. Med. Chem. Lett., 13*, 1813–1815.
21. Deeks, S. G., Smith, M., Holodniy, M., & Kahn, J. O., (1997). HIV-1 protease inhibitors: A review for clinicians. *JAMA, 277*(2), 145–153.
22. Tewtrakul, S., Subhadhirasakul, S., Puripattanavong, J., & Panphadung, T., (2003). HIV-1 protease inhibitory substances from the rhizomes of *Boesenbergia pandurata* holtt. *Songklanakarin. J. Sci. Technol., 25*(4), 503–508.
23. Duintjer, T. R. J., Pallansch, M. A., Chumakov, K. M., Halsey, N. A., Hovi, T., Minor, P. D., Modlin, J. F., et al., (2013). Expert review on poliovirus immunity and transmission. *Risk Analysis, 33*(4), 544–605.
24. Ahmed, A. M., Sadiq, G. T., Kareem, N. H., & Abdullatif, M. O., (2013). Inhibitory effects of chalcone on the replication of polio virus In Vitro. *J. Baghdad Sci., 10*(2), 319–322.
25. Patick, A. K., (2006). Rhinovirus chemotherapy. Antiviral research., 71(2, 3), 391–396.
26. Ishitsuka, H., Ninomiya, Y. T., Ohsawa, C., Fujiu, M., & Suhara, Y., (1982). Direct and specific inactivation of rhinovirus by chalcone Ro 09-0410. *Antimicrob. Agents Chemother., 22*(4), 617–621.
27. Yin, Z., Patel, S. J., Wang, W. L., Wang, G., Chan, W. L., Rao, K. R., Alam, J., et al., (2006). Peptide inhibitors of dengue virus NS3 protease. Part 1: Warhead. *Bioorganic and Medicinal Chemistry Letters, 16*(1), 36–39.

28. Kiat, T. S., Pippen, R., Yusof, R., Ibrahim, H., Khalid, N., & Rahman, N. A., (2006). Inhibitory activity of cyclohexenyl chalcone derivatives and flavonoids of finger root, *Boesenbergia rotunda* (L.), towards dengue-2 virus NS3 protease. *Bioorg. Med. Chem. Lett. 16*, 3337–3340.
29. Mahapatra, D. K., (2014). Neuraminidase inhibitors for effective treatment of influenza. *Int. J. Pharm. Res. Technol., 4*(1), 22–31.
30. Dao, T., Tung, B., Nguyen, P., Thuong, P., Yoo, S., Kim, E., Kim, S., & Oh, W., (2010). C-Methylated flavonoids from *Cleistocalyx operculatus* and their inhibitory effects on novel influenza A (H1N1) neuraminidase. *J. Nat. Prod. 73*, 1636–1642.
31. Dao, T. T., Nguyen, P. H., Lee, H. S., Kim, E., Park, J., Lim, S. I., & Oh, W. K., (2011). Chalcones as novel influenza A (H1N1) neuraminidase inhibitors from glycyrrhiza inflate. *Bioorg. Med. Chem. Lett. 21*, 294–298.
32. Park, J., Jeong, H. J., Kim, Y. M., Park, S., Rho, M., Park, K. H., Ryu, Y. B., & Lee, W. S., (2011). Characteristic of alkylated chalcones from *Angelica keiskei* on influenza virus neuraminidase inhibition. *Bioorg. Med. Chem. Lett., 21*, 5602–5604.
33. Scholthof, K. B., (2004). Tobacco mosaic virus: A model system for plant biology. *Annu. Rev. Phytopathol, 42*, 13–34.
34. Ye, Y., Dong, W., Liu, C., et al., (2016). Chalcones from *Desmodium podocarpum* and their anti-tobacco mosaic virus activity. *Chem. Nat. Compd., 52*(3), 409–412.

CHAPTER 9

Ethnoscience, Alternative Medicine, Inflammation, and Pain Models

FRANCISCO TORRENS[1] and GLORIA CASTELLANO[2]

[1]*Institut Universitari de Ciència Molecular, Universitat de València, Edifici d'Instituts de Paterna, P.O. Box 22085, E-46071 València, Spain, E-mail: torrens@uv.es*

[2]*Departamento de Ciencias Experimentales y Matemáticas, Facultad de Veterinaria y Ciencias Experimentales, Universidad Católica de Valencia San Vicente Mártir, Guillem de Castro-94, E-46001 València, Spain*

ABSTRACT

Plants form the basis of African traditional medicine, which contributed significantly to the reduction of mortality, morbidity, and disability owing to many diseases. Medicinal plants were reported to be an essential aspect of oral health delivery cutting across all facets of oral healthcare. One should avoid the medical practice that carries any supplementary description since he will surely be paying expensive for something that no one can guarantee people to work. Among the therapies are well-known ones (homeopathy, acupuncture), and others more strange that somebody invents and applies them. Though traditional healers are trained in handling plant material, additional training is required for plants to be safe for medicine. Medicinal plants present benefits in the management of oral diseases but face challenges. Research on medicinal plants used in African traditional medicine in oral health is scanty. Mediterranean Diet Pyramid is the basis for integrative medicine for veterans with disabilities, but patient-centered and interprofessional approaches represent the real added value for the best healthcare management.

9.1. INTRODUCTION

Setting the scene: ethnopharmacology (EP) studies for drug development with special reference to Asteraceae, recommendations, research gaps, pain, pleasure, pelvic floor (PF), the identification of alert signs in PF problems, and the rescue of the noradrenergic system as a novel pharmacological strategy in the treatment of chronic pain (CP) with focus on microglia activation. One should avoid the medical practice that carries any description since he will surely be paying expensive for something that no one can guarantee people to work. Among the therapies are well-known ones, and others more exotic that somebody invents and applies them. Though traditional medicine (TM) healers are trained in handling medicinal plants (MPs) material, training is required for MPs to be safe for medicine. The MPs present benefits in oral-diseases management but face challenges. Research on MPs used in African TM (ATM) in oral health is incomplete. A monograph on MPs for holistic healing was published [1]. The MPs played, and continue to play, a major role in the lives of people worldwide [2]. To get the integration of the ethnosciences in Social normality requires a teachers' model [3]. It was informed bioactive (BA) phenolics of *Artemisia*, the chromatomass-spectrometric profile of Siberian species, and inhibitory potential *vs.* a-amylase and a-glucosidase [4]. A review identified MPs used by ATM healers in preventive, therapeutic, and surgical aspects in oral health, and recognized research gaps [5]. Important MP families in Thailand were reported [6]. The therapeutic use–mode of action gap in remedial herbs was closed [7].

The sesquiterpene lactone (STL) γ-lactone budlein A (*cf.* Figure 9.1), from *Viguiera robusta*, inhibits antigen-induced arthritis in mice with a role of nuclear factor κ-light-chain-enhancer of activated B cells (NF-κB) and cytokines [8]. It alleviates pain and inflammation in an acute gout arthritis model in mice [9]. Cannabinoids and pain were reviewed with insights from old molecules [10]. Neuropharmacology of dementia neuropsychiatric symptoms and the role of pain were revised [11]. Modulation of the negative affective dimension of pain was discussed, with a focus on selected neuropeptidergic system contributions [12]. The rescue of the noradrenergic system was examined as a novel pharmacological strategy in CP treatment [13].

Earlier publications in *Nereis*, etc. classified yams [14], lactic acid bacteria [15], fruits [16], spices [17], and oil legumes [18] by principal component, cluster, and meta-analysis. The molecular classifications of phenolic compounds derived from the cinnamic and benzoic acids from *Posidonia oceanica* [19], flavonoids [20], stilbenoids [21], triterpenoids, steroids from *Ganoderma* [22], isoflavonoids from *Dalbergia parviflora* [23], STLs [24,25] and artemisinin (ART) derivatives [26] were informed. A tool for interrogation of the macromolecular structure was reported [27]. Mucoadhesive polymer hyaluronan favors transdermal penetration absorption of caffeine [28,29]. Polyphenolic phytochemicals in cancer prevention, therapy, bioavailability, and bioefficacy were reviewed [30]. The natural product (NP) antioxidants (AOs) from herbs and spices improved the oxidative stability (OS) of vegetable oils (VOs) [31]. The VO composition–OS relationship was revealed *via* a multifactorial approach [32]. It was informed chemical and biological screening approaches to phytopharmaceuticals [33], cultural interbreeding in indigenous and scientific EP [34], MPs ethnobotany (EB) studies, underutilized wild edible plants, food, medicine [35], biodiversity (BD) as a source of drugs, *Cordia*, *Echinacea*, *Tabernaemontana*, *Aloe* [36], immunomodulatory molecules from Himalayan MPs [37], antimicrobial, antioxidant, the composition of *Verbena carolina*, *Mentha* [38], phylogenesis by information entropy, avian birds, influenza [39], pain and pleasure [40] The purpose of the present work is to review EP studies for the development of drugs with special reference to Asteraceae, recommendations, research gaps, pain, pleasure, PF, alert-signs identification in PF problems and the rescue of the noradrenergic system as a pharmacological strategy in CP treatment with a focus on microglia activation. The aim of this study is to initiate a debate suggesting a number of questions (Q) that can arise when addressing subjects of pain, pleasure, how pain is avoidable, PF, pregnancy as a PF's crucial moment, anal incontinence and proctalgia understanding problem improving life quality, in different fields and providing, when possible, hypotheses (H) on the integral theory of female PF, revolution in diagnosis and treatment.

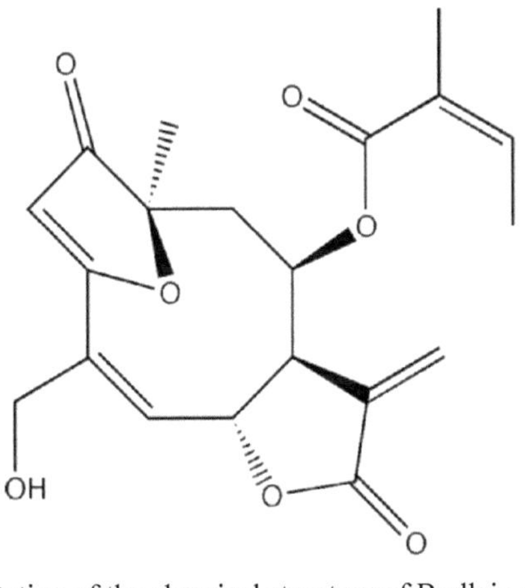

FIGURE 9.1 Representation of the chemical structure of Budlein A.

9.2 ETHNOPHARMACOLOGICAL STUDIES FOR THE DEVELOPMENT OF DRUGS

One of the most important genera of Asteraceae is probably *Artemisia*, as it is used worldwide in TM as a source for spices (e.g., *A. dracunculus*), ingredient for liquors (e.g., *A. absinthium*) and source for ART [41]. The discovery of ART fundamentally changed the treatment of malaria around the globe and led to the awarding of the Nobel Prize in Physiology or Medicine 2015.

9.3 DISCUSSION ON ETHNOPHARMACOLOGY

The EP forms the basis for the development of ethnic-groups-TM-based therapeutics. Local-knowledge preservation, indigenous-medical-systems promotion in primary healthcare (HC), and BD conservation became a concern. Phytochemical (PC) analysis (PCA) allowed an examination and isolation of BA natural products (NPs), and their identification and structure elucidation. Herbal medicines (HMs) from TM received attention in the prevention and treatment of chronic metabolic disorders, for example, diabetes mellitus (DM). Species of MPs are anti-DM, for example, MP extracts that inhibit carbohydrate (CH_2O) hydrolysis enzymes (α-amylase, α-glucosidase), which play a role in CH_2O digestion. The enzymes inhibition averts DM. From the scientific-method application, one defines *medicine* as *that based on therapies or treatments that have proven valid in clinical trials or controlled experiments*, from which medicine does not require *surnames* [42]. Talking about *alternative* (AM), complementary, integrative, Chinese, etc. *medicines*, one suspects. The problem: They use therapies that do not present scientific evidence that supports them. If they had it, they would no longer be AM but *medicine*; for example, treatments existed that were applied in Chinese TM (TCM), shown to be effective and incorporated in the usual therapies.

Plants as the basis of ATMs were used for thousands of years and are in use. Plant-based TMs play a role in HC systems in regions, especially Africa, where modern drugs are not affordable. The TM based on MPs is used by the *pharma* for drug discovery for years before the chemical synthesis. A 25% of drugs are directly derived from plants, while others are formed from prototype compounds isolated from plant species. Reliability and usage of HM NPs became important because of the adverse effects (AEs) and complications of synthetic drugs (SDs): need to identify and recognize MPs beneficial effects. Using MP-based drugs reported being higher in developing countries because modern life-saving drugs are beyond reach, despite spending half of their total gross domestic product (GDP) on drugs and HC, which alternative constitutes a drug financial burden reduction strategy. The need for alternative prevention and treatment options for oral diseases that be safe, effective and economical was deemed necessary for reasons [e.g., high oral disease incidence, risen resistance pathogenic bacteria to antibiotics and chemotherapeutics (CT), opportunistic infections in immunocompromised individuals, financial difficulties in developing countries]. Bacterial resistance to most antibiotics used to treat oral infections (penicillins, cephalosporins, erythromycin, tetracycline, metronidazole derivatives) was reported. In antibacterial function, the drugs alter oral microbiota and exert AEs (e.g., vomiting, diarrhea, tooth staining). Using TM/MPs in developing countries as the normative basis for the maintenance of health was advocated. The TMs were used to treat orofacial problems in Asia for more than 2,000y, and shown to be effective in recurrent aphthous stomatitis, oral lichen planus, leukoplakia, and Sjögren's syndrome. However, the treatment lacked standard criteria of post-treatment assessment and laboratory evidence. The TCM and naturopathic medicine reduced pain and psychosocial interference in temporomandibular disorders.

The MPs were used for medicinal purposes since 60,000y. Millions depend on MPs for their well-being. In the tropics, MPs are used on a regular basis in rural communities where drugs are hard to obtain or unavailable, which is in contrast to westernized societies where MPs are typically used as an alternative or supplement to prescribed drugs. The MPs are important for people not only as a primary source of drugs but also as PC building blocks for drug development. A 67% of drugs used in CT are derived from NPs, which applies to the discoveries of active compounds (e.g., vincristine, taxol, ART). The MPs generate a cash income for rural dwellers. Factors influence how people select MPs for medicinal purposes (tradition, efficacy, abundance, accessibility, the doctrine of signatures, taxonomic affiliation). Plant leaves, roots, barks, and extracts were used since the dawn of man to treat discomforts and maladies. Remedial-HM healing properties were identified *via* a trial-and-error learning process, which was passed *via* generations. The TMs represent an established therapeutic alternative to SDs in parts of the world. However, a lack of understanding exists about the specific chemical ingredients and exact mechanisms of action, by which MPs exert their therapeutic effect.

9.4 RECOMMENDATIONS

The MPs present enormous benefits in the management of oral diseases but face challenges (e.g., toxicity, processing, follow-ups). Research on MPs used in ATM in oral health is scanty and incomplete. Identified research gaps follow and they recommended carrying out research in the areas.

9.5 RESEARCH GAPS

1. The PCA of MPs used in Africa. Plants used for the treatment of oral cancers were not reported, which area needs to be exploited.
2. Processing, storage, and preservation of MPs need to be addressed. Further research is needed in the area.
3. Literature concerning the use of TMs in maxillofacial injuries, for example, jaw fractures, is lacking.
4. Development of potential bioproducts (e.g., HM toothpaste, mouthwashes, and gaggles, toothpowder) to treat diseases of the oral cavity, which may be more cost-effective and present fewer AEs than commercial brands already in the market.
5. No international research platform exists to address gaps in MPs scientific knowledge, which may be useful in the management of oral and dental diseases.
6. Capacity building and training opportunities in EB, particularly in regions with gaps in such resources.
7. Effect of the toothbrush chewing sticks, *vs.* other common bacteria that are associated with dental diseases, was not studied.

9.6 PAIN AND PLEASURE

Pain may be identified as being a complex and varied set of perceptions, which are characterized by an unpleasant and tense nature [43]. It functions as a warning sign and triggering defense and preservation reactions, presenting an important defense function. It, at its upper limit, should not interfere with the normal activities of the individual, but cannot be unnoticed that it be not felt before serious damage occurs to tissues. It can be categorized into two types: neuropathic and nociceptive. Neuropathic pain is not related to direct threats to the integrity of tissues, but it originates from previous lesions that led to the dysfunction of the nerve pathways (e.g., in the synapses, inhibitory mechanisms). Nociceptive pain responds to mechanical, thermal, electrical, and chemical nociceptive stimuli usually associated with aggression. Regarding the chemical stimuli, the mechanisms *via* which they induce the emergence of a difference in membrane potential, in nociceptors, generally involve their interaction with ionic channels that are sensitive to voltage, with receptors associated with tyrosine kinase or G-protein, which is the most common. The mediators that resort to the mechanism presents specific receptors, which are coupled to G-proteins that differ in the α-subunit, activating effector pathways. Among the endogenous substances, the one that is the most potent to induce pain is bradykinin. On the other hand, the stimulation of the primary afferent nociceptors results in the release of neuropeptides (i.e., substance-P, neurokinin-A), which contribute to a greater stimulation of the inflammatory process.

9.7 PELVIC FLOOR: IDENTIFICATION OF ALERT SIGNS IN PELVIC-FLOOR PROBLEMS

De Andrés organized Day *Pain Is Avoidable: Pelvic Floor* [44]. Some questions were raised.

Q1. Is it normal that a woman has a loss of urine?
Q2. Is man free from pelvic problems?

In Round Table 1 on alert-signs identification in PF problems, García Oms proposed Q/H on pregnancy as a crucial moment for PF [45].

Q3. What is PF?

H1. (Petros, 1990). *Integral Theory of the Female* PF: *A Revolution in Diagnosis//Treatment* [46].
Q4. What can we do PF pregnancy-labor?

Roig raised Q on anal incontinence/proctalgia understanding problem improving life [47].

Q5. How to measure incontinence?
Q6. What to do?
Q7. What not to do?

9.8 NORADRENERGIC-SYSTEM RESCUE AS PHARMACOLOGY IN CHRONIC PAIN

Different types of pain evolve to a chronic condition characterized by hyperalgesia and allodynia, with an abnormal response to normal or even innocuous stimuli, respectively. A role in endogenous analgesia is recognized to descending noradrenergic pathways, which originate from the *locus coeruleus* and project to the dorsal horn of the spinal cord, in which system impairment is associated with pain chronicization. Activation of glial cells, in particular microglia, to a pro-inflammatory state was implicated in the transition from acute to CP. α2- and β2-Adrenergic receptors are expressed in microglia, and activation leads to the acquisition of an anti-inflammatory phenotype. Descending noradrenergic system–neuroinflammation interconnection was reviewed, focusing on drugs that, rescuing the noradrenergic control, exert an anti-inflammatory effect, leading to analgesia [48]. The potential efficacy in the treatment of neuropathic pain of different drugs was analyzed. On one side, drugs acting as inhibitors of the reuptake of serotonin (5-hydroxytryptamine, 5-HT) and noradrenaline (NA) (e.g., duloxetine, venlafaxine) and, on the other, tapentadol, inhibitor of the reuptake of NA and agonist of μ-opioid receptor (MOR).

9.9 DISCUSSION ON PAIN

Mediterranean Diet Pyramid (MDP) is the basis for integrative medicine for veterans with disabilities, but patient-centered and interprofessional approaches represent the real added value for the best HC management [49]. A comprehensive approach should include physical medicine and rehabilitation clinicians, pharmacists, and nutritionist to prevent malnutrition, self-prescription of complementary and alternative medicine (CAM), and food– and/or nutraceutical–drug interactions according to a biopsychosocial model. Personalized HC for chronic non-communicable diseases that impact the quality of life should consider gut microbiota, genetic and epigenetic factors, and moods and hormones involved in stress response rather than functional status. In order to avoid potential food-drug interactions, plant foods should be chosen in those containing low phytochemicals and high micronutrients. If this is difficult to reach, vitamin (Vit) and/or mineral supplementation is recommended (top of MDP for veterans).

9.10 FINAL REMARKS

From the present results and discussion, the following final remarks can be drawn.

1. One should avoid the medical practice that carries any additional or supplementary description since he will surely be paying expensive for something that no one can guarantee people to work. Among the therapies are well-known ones, for example, homeopathy or acupuncture, and others more strange or exotic that somebody invents and applies them.
2. Though traditional healers are trained in handling plant material, additional training is required for plants to be safe for medicinal uses.
3. Medicinal plants present enormous benefits in the management of oral diseases but face challenges, for example, toxicity, processing, and follow-ups. Research on medicinal plants used in African traditional medicine in oral health is scanty and incomplete.
4. Mediterranean Diet Pyramid is the basis for integrative medicine for veterans with disabilities, but patient-centered and interprofessional approaches represent the real added value for the best healthcare management. A comprehensive approach should include physical medicine and rehabilitation clinicians, pharmacists, and nutritionist to prevent malnutrition, self-prescription of complementary and alternative medicine, and food- and/or nutraceutical-drug interactions, according to a biopsychosocial model. Personalized healthcare for chronic non-communicable diseases that impact the quality of life should consider gut microbiota, genetic and epigenetic factors, and moods and hormones involved in stress response rather than functional status. In order to avoid potential food-drug interactions, plant foods should be chosen in those containing low phytochemicals and high micronutrients. If this is difficult to reach, vitamin and/or mineral supplementation is recommended (top of Mediterranean Diet Pyramid for veterans).

ACKNOWLEDGMENTS

The authors thanks the support from Fundacion Universidad Catolica de Valencia San Vicente Martir (Project No. 2019-217-001UCV).

KEYWORDS

- **African traditional medicine**
- **biochemical mechanism**
- **dental caries**
- **experimental technique**
- **neuroinflammation**
- **neuropathic pain**
- **oral health**

REFERENCES

1. Lall, N., (2018). *Medicinal Plants for Holistic Healing*. Elsevier Academic: London, UK.
2. McGaw, L. J., Srivastava, A. K., Lin, C. H., & Steenkamp, V., (2019). Book review: Medicinal plants for holistic healing. *Front. Pharmacol., 10*, 1053–1–2.
3. Climent, D., (2018). An educating region and an educational proposal: Ethnoscience (I). *DAUALDEU, 2018*(15), 5–7.
4. Olennikov, D. N., Chirikova, N. K., Kashchenko, N. I., Nikolaev, V. M., Kim, S. W., & Vennos, C., (2018). Bioactive phenolics of the genus *Artemisia* (Asteraceae): HPLC-DAD-ESI-TQ-MS/MS profile of the Siberian species and their inhibitory potential against a-amylase and a-glucosidase. *Front. Pharmacol., 9*, 756–1–27.
5. Agbor, A. M., & Naidoo, S., (2019). Plants used by African traditional healers in the management of oral diseases: A review. *Int. J. Res.-Granthaalayah, 7*(8), 273–286.
6. Phumthum, M., Balslev, H., & Barfod, A. S., (2019). Important medicinal plant families in Thailand. *Front. Pharmacol., 10*, 1125–1–14.
7. Olivés, J., & Mestres, J., (2019). Closing the gap between therapeutic use and mode of action in remedial herbs. *Front. Pharmacol., 10*, 1132–1–11.
8. Zarpelon, A. C., Fattori, V., Souto, F. O., Pinto, L. G., Pinho-Ribeiro, F. A., Ruiz-Miyazawa, K. W., Turato, W. M., et al., (2017). The sesquiterpene lactone, budlein A, inhibits antigen-induced arthritis in mice: Role of NF-κB and cytokines. *Inflammation, 40*, 2020–2032.
9. Fattori, V., Zarpelon, A. C., Staurengo-Ferrari, L., Borghi, S. M., Zaninelli, T. H., Da Costa, F. B., Alves-Filho, J. C., et al., (2018). a sesquiterpene lactone from *Viguiera robusta*, alleviates pain and inflammation in a model of acute gout arthritis in mice. *Front. Pharmacol., 9*, 1076–1–13.
10. Vuckovic, S., Srebro, D., Vujovic, K. S., Vucetic, C., & Prostran, M., (2018). Cannabinoids and pain: New insights from old molecules. *Front. Pharmacol., 9*, 1259–1–19.
11. Scuteri, D., Rombolà, L., Morrone, L. A., Bagetta, G., Sakurada, S., Sakurada, T., Tonin, P., & Corasaniti, M. T., (2019). Neuropharmacology of the neuropsychiatric symptoms

of dementia and role of pain: Essential oil of bergamot as a novel therapeutic approach. *Int. J. Mol. Sci., 20*, 3327–1–13.
12. Caputi, F. F., Rullo, L., Stamatakos, S., Candeletti, S., & Romualdi, P., (2019). Modulation of the negative affective dimension of pain: Focus on selected neuropeptidergic system contributions. *Int. J. Mol. Sci., 20*, 4010-1–13.
13. Caraci, F., Merlo, S., Drago, F., Caruso, G., Parenti, C., & Sortino, M. A., (2019). Rescue of noradrenergic system as a novel pharmacological strategy in the treatment of chronic pain: Focus on microglia activation. *Front. Pharmacol., 10*, 1024–1–8.
14. Torrens-Zaragozá, F., (2013). Molecular categorization of yams by principal component and cluster analyses. *Nereis, 2013*(5), 41–51.
15. Torrens-Zaragozá, F., (2014). Classification of lactic acid bacteria against cytokine immune modulation. *Nereis, 2014*(6), 27–37.
16. Torrens-Zaragozá, F., (2015). Classification of fruits proximate and mineral content: Principal component, cluster, meta-analyses. *Nereis, 2015*(7), 39–50.
17. Torrens-Zaragozá, F., (2016). Classification of food spices by proximate content: Principal component, cluster, meta-analyses, *Nereis, 2016*(8), 23–33.
18. Torrens, F., & Castellano, G., (2014). From Asia to Mediterranean: Soya bean, Spanish legumes and commercial *soya bean* principal component, cluster and meta-analyses. *J. Nutr. Food Sci., 4*(5), 98.
19. Castellano, G., Tena, J., & Torrens, F., (2012). Classification of polyphenolic compounds by chemical structural indicators and its relation to antioxidant properties of *Posidonia oceanica* (L.) Delile. *MATCH Commun. Math. Comput. Chem., 67*, 231–250.
20. Castellano, G., González-Santander, J. L., Lara, A., & Torrens, F., (2013). Classification of flavonoid compounds by using entropy of information theory. *Phytochemistry, 93*, 182–191.
21. Castellano, G., Lara, A., & Torrens, F., (2014). Classification of stilbenoid compounds by entropy of artificial intelligence. *Phytochemistry, 97*, 62–69.
22. Castellano, G., & Torrens, F., (2015). Information entropy-based classification of triterpenoids and steroids from *Ganoderma*. *Phytochemistry, 116*, 305–313.
23. Castellano, G., & Torrens, F., (2015). Quantitative structure–antioxidant activity models of isoflavonoids: A theoretical study. *Int. J. Mol. Sci., 16*, 12891–12906.
24. Castellano, G., Redondo, L., & Torrens, F., (2017). QSAR of natural sesquiterpene lactones as inhibitors of Myb-dependent gene expression. *Curr. Top. Med. Chem., 17*, 3256–3268.
25. Torrens, F., & Castellano, G., (2020). Structure–activity relationships of cytotoxic lactones as inhibitors and mechanisms of action. *Curr. Drug Discov. Technol* (Vol. 17 pp. 166-182) in press.
26. Torrens, F., Redondo, L., & Castellano, G., (2017). Artemisinin: Tentative mechanism of action and resistance. *Pharmaceuticals, 10*, 20–4–4.
27. Torrens, F., & Castellano, G., (2014). A tool for interrogation of macromolecular structure. *J. Mater. Sci. Eng. B, 4*(2), 55–63.
28. Torrens, F., & Castellano, G., (2014). Mucoadhesive polymer hyaluronan as biodegradable cationic/zwitterionic-drug delivery vehicle. *ADMET DMPK, 2*, 235–247.
29. Torrens, F., & Castellano, G., (2015). Computational study of nanosized drug delivery from *cyclo*dextrins, crown ethers and hyaluronan in pharmaceutical formulations. *Curr. Top. Med. Chem., 15*, 1901–1913.

30. Estrela, J. M., Mena, S., Obrador, E., Benlloch, M., Castellano, G., Salvador, R., & Dellinger, R. W., (2017). Polyphenolic phytochemicals in cancer prevention and therapy: Bioavailability *versus* bioefficacy. *J. Med. Chem., 60*, 9413–9436.
31. Redondo-Cuevas, L., Castellano, G., & Raikos, V., (2017). Natural antioxidants from herbs and spices improve the oxidative stability and frying performance of vegetable oils. *Int. J. Food Sci. Technol., 52*, 2422–2428.
32. Redondo-Cuevas, L., Castellano, G., Torrens, F., & Raikos, V., (2018). Revealing the relationship between vegetable oil composition and oxidative stability: A multi-factorial approach. *J. Food Compos. Anal., 66*, 221–229.
33. Torrens, F., & Castellano, G., (2020). Chemical and biological screening approaches to phytopharmaceuticals. In: Pourhashemi, A., Deka, S. C., & Haghi, A. K., (eds.), *Research Methods and Applications in Chemical and Biological Engineering* (pp. 3–12). Apple Academic–CRC: Waretown, NJ.
34. Torrens, F., & Castellano, G., (2020). Cultural interbreeding in indigenous/scientific ethnopharmacology. In: Pourhashemi, A., Deka, S. C., & Haghi, A. K., (eds.), *Research Methods and Applications in Chemical and Biological Engineering* (pp. 25–35). Apple Academic–CRC: Waretown, NJ.
35. Torrens, F., & Castellano, G., (2020). Molecular Chemistry and Biomolecular Engineering: Integrating Theory and Research with Practice. In: Pogliani, L., Torrens, F., & Haghi, A. K., (eds.), *Innovations in Physical Chemistry* (pp. 63–71). Apple Academic–CRC: Waretown, NJ.
36. Torrens, F., & Castellano, G., (2020). Biodiversity as a source of drugs: Cordia, Echinacea, tabernaemontana and Aloe. In: Aguilar, C. N., Ameta, S. C., & Haghi, A. K., (eds.), *Green Chemistry and Biodiversity. Principles, Techniques, and Correlations* (pp. 9–21). Apple Academic–CRC: Waretown, NJ.
37. Torrens, F., & Castellano, G., (2020). Immunomodulatory molecules from Himalayan medicinal plants. In: Esteso, M. A., Ribeiro. A. C. F., & Haghi, A. K., (eds.), *Chemistry and Chemical Engineering for Sustainable Development: Best Practices and Research Directions*. Apple Academic–CRC: Waretown, NJ.
38. Torrens, F., & Castellano, G. (2020). Antimicrobial, Antioxidant, and composition of verbena carolina and *Mentha*. In: Mahapatra, D. K., Talele, S. G., & Haghi, A. K., (eds.), *Applied Pharmaceutical Science and Microbiology: Novel Green Chemistry Methods and Natural Products*. Apple Academic–CRC: Waretown, NJ.
39. Torrens, F., & Castellano, G., (2008). *Phylogenesis by Information Entropy: Avian Birds and (1918).* Influenza virus. In: Turner, S., & Yunus, J., (eds.), *Modeling and Simulation* (Vol. 2; pp. 1, 2). IEEE: London.
40. Torrens, F., & Castellano, G., (2020). Pain and pleasure. In: Pogliani, L., Torrens, F., & Haghi, A. K., (eds.), *Molecular Chemistry and Biomolecular Engineering: Integrating Theory and Research with Practice* (pp. 173–190). Apple Academic–CRC: Waretown, NJ.
41. Panda, S. K., Da Silva, L. C. N., Sahal, D., & Leonti, M., (2019). Editorial: Ethno pharmacological studies for the development of drugs with special reference to *Asteraceae*. *Front. Farmacol., 10*, 955–1–2.
42. Mulet, J. M., (2018). Exotic therapies. *DAUALDEU 2018*(14), 4.
43. Santos, T., Ferreira, J., Castel-Branco, M., & Figueiredo, I. V., (2019). The use of animal models of inflammation and pain in biomedical research. In: Vakhrushev, A. V., Haghi, R., & De Julián-Ortiz, J. V., (eds.), *Physical Chemistry for Chemists and Chemical*

Engineers: Multidisciplinary Research Perspectives (pp. 11–46). Apple Academic–CRC: Waretown, NJ.
44. De Andrés, J., (2018). Book of Abstracts, XIV Annual Conference Pain Is Avoidable: Pelvic Floor, València, Spain. Fundación Valenciana de Estudis Avanzados: València, Spain.
45. García, O. J., (2018). Book of Abstracts, XIV Annual Conference Pain Is Avoidable: Pelvic Floor, València, Spain. Fundación Valenciana de Estudis Avanzados: València, Spain, 2018; RT-1.
46. Petros, P. E. P., & Ulmsten, U. I., (1990). An integral theory of female urinary incontinence: Experimental and clinical considerations. *Acta Obstet. Gynecol. Scand., 69*(Suppl. 153), 7–31.
47. Roig, J. V., (2018). Book of Abstracts, XIV Annual Conference Pain Is Avoidable: Pelvic Floor, València, Spain, Fundación Valenciana de Estudis Avanzados: València. Spain, 2018; RT-1.
48. Caraci, F., Merlo, S., Drago, F., Caruso, G., Parenti, C., & Sortino, M. A., (2019). Rescue of noradrenergic system as a novel pharmacological strategy in the treatment of chronic pain: Focus on microglia activation. *Front. Pharmacol., 10*, 1024–1–8.
49. Sciarra, T., Ciccotti, M., Aiello, P., Minosi, P., Munzi, D., Buccolieri, C., Peluso, I., et al., (2019). Polypharmacy and nutraceuticals in veterans: Pros and cons. *Front. Pharmacol., 10*, 994–1–5.

CHAPTER 10

Orange Fruit and Bioplastics with Limonene and No Bisphenol A

FRANCISCO TORRENS[1] and GLORIA CASTELLANO[2]

[1]*Institut Universitari de Ciència Molecular, Universitat de València, Edifici d'Instituts de Paterna, P.O. Box 22085, E-46071 València, Spain, E-mail: torrens@uv.es*

[2]*Departamento de Ciencias Experimentales y Matemáticas, Facultad de Veterinaria y Ciencias Experimentales, Universidad Católica de Valencia San Vicente Mártir, Guillem de Castro-94, E-46001 València, Spain*

ABSTRACT

Results sketch an evolutionary outline for citrus, a situation that confronts present taxonomic and phylogenetic views, and indicates a reformulation of the genus *Citrus*. Chemistry builds up a viaduct between the recognizable and the basic. This account gives a wide-ranging insight of such a flyover but, with any luck, one can see across the mist and comprehend how a chemist reason. This explanation will explain to people that the world is, at the same time, simple, complicated, and complex. It is complex because even a leaf is composed of countless kinds of substances with a diversity that changes with the seasons. It is simple because scientists can unravel the leaf, distinguish its components, and see how in summer one overlooks and another, in autumn. The job for disentanglement is partial, and an era can pass before being acquainted with certainly all particulars about the dissimilarities between an apple and an orange, much less between a woman and a man. Notwithstanding, this report explains to people that the throng of compounds that make up the world is every timeless an amorphous worm, and a huge collection of comprehensible particularized characters.

10.1 INTRODUCTION

Setting the scene: some components of an orange fruit and bioplastics with limonene and no bisphenol A. Findings draw an evolutionary framework for citrus, a scenario that challenges current taxonomic and phylogenetic thoughts, and points to a reformulation of the genus *Citrus*. Chemistry establishes a bridge between the familiar and the fundamental. This report provides a general view of such a bridge but, with a bit of luck, one can see across the fog and understand how a chemist thinks. This description will teach people that the world is both simple and complicated. It is complicated because even a leaf is composed of an innumerable type of molecules with a variety that changes with the seasons. It is simple because scientists can unravel the leaf, distinguish its components, and see how in summer one of its components dominates and another, in autumn. The task for unraveling is far from complete, and an era can pass before knowing with certainty all the details about the differences between an orange and a banana, much less between a man and a woman. Nevertheless, this account teaches people that the crowd of molecules that make up the world is each timeless an amorphous worm, and ever more a group of understandable individual personalities.

The identification of three kinds of Citri Reticulatae Pericarpium was informed, based on deoxyribonucleic acid barcoding and high-performance liquid chromatography (HPLC)-diode array detection–electrospray ionization (ESI)/mass spectrometry (MS)/MS combined with chemometric analysis [1]. A review identified medicinal plants used by traditional healers in preventive, therapeutic, and surgical aspects in oral health, and recognized some research gaps, which further studies should be carried out [2]. Analysis of flavonoid metabolites was reported in citrus peels (*Citrus reticulata* Dahongpao) *via* ultra-HPLC-ESI-MS/MS [3]. Bigger data approach was informed to the analysis of essential oils (EOs) and antifungal activity *vs. Aspergillus niger, Candida albicans,* and *Cryptococcus neoformans* [4]. The gap between therapeutic use and mode of action in remedial herbs was closed [5].

In earlier publications in *Nereis*, etc., it was reported the molecular classification of yams [6], soya bean, Spanish legumes, commercial *soya bean* [7, 8], fruits proximate and mineral content [9], and food spices proximate content [10] by principal component, cluster, and meta-analyses. Polyphenols [11], flavonoids [12], stilbenoids [13], triterpenoids, steroids [14], isoflavonoids [15], natural sesquiterpene lactones [16, 17], artemisinin and their derivatives [18–21] were analyzed. It was informed the classification of *Citrus*, principal components, cluster, and meta-analyses [22], the extraction of natural products found in vegetal species clove and citrus [23], the world of biological activities and safety of *Citrus* spp. EOs [24], *Citrus* origin, evolution, postharvest, heat, p*H*, filtration, transgenic [25], *Citrus* β-cryptoxanthyn reduces 30% body fat, and nobiletin benefits [26]. The main aim of the present report is to review some components of an orange fruit and bioplastics with limonene and no bisphenol-A.

10.2 SOME COMPONENTS OF AN ORANGE FRUIT

In some substances, one can show the molecules that *typify* their characteristics [27]. Although one cannot show all components of an orange fruit, he can do it with the typical molecules that contribute to its taste or texture. People manage to start distinguishing molecularly an orange from a banana fruit. The odor of orange fruits is because of a terpene molecule (D-limonene, *cf.* Figure 10.1a), which is the mirror image of the molecule that contributes to the odor of lemon fruits (L-limonene, Figure 10.1b).

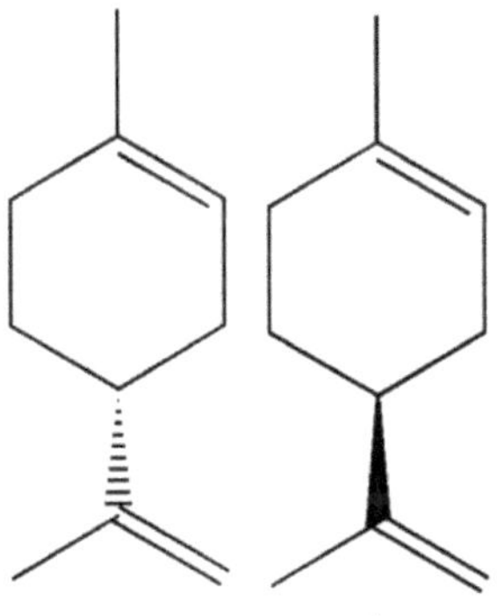

a b

FIGURE 10.1 Molecular structures of (a) D-limonene and (b) L-limonene.

The color of the orange fruit is because of anthocyanin molecules (*cf.* Figure 10.2).

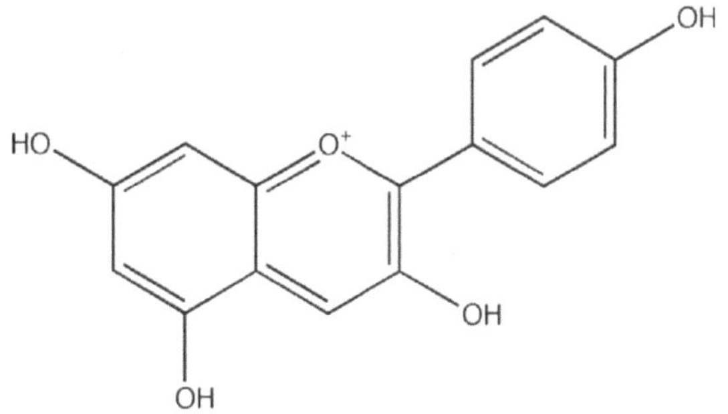

FIGURE 10.2 Molecular structure of pelargonidin, the simplest anthocyanidin.

The acidity of the orange fruit is because of citric acid (*cf.* Figure 10.3).

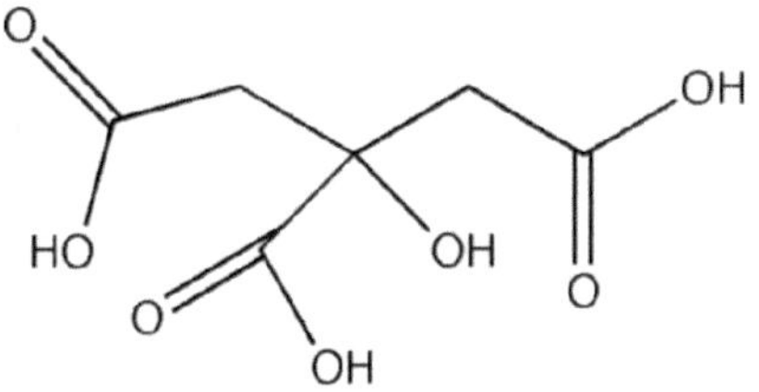

FIGURE 10.3 Molecular structure of citric acid.

10.3 BIOPLASTICS WITH LIMONENE AND NO BISPHENOL A

One can find plastic polycarbonates everywhere from a cell phone sleeve to an aircraft window, which explains that tons of the material exist all over the world [28]. However, one of the drawbacks of the materials is that bisphenol-A (BPA) is involved in its preparation, which, according to studies, is harmful to health and considered by European Chemicals Agency (ECHA) as a worrying product (*cf.* Figure 10.4) [29, 30].

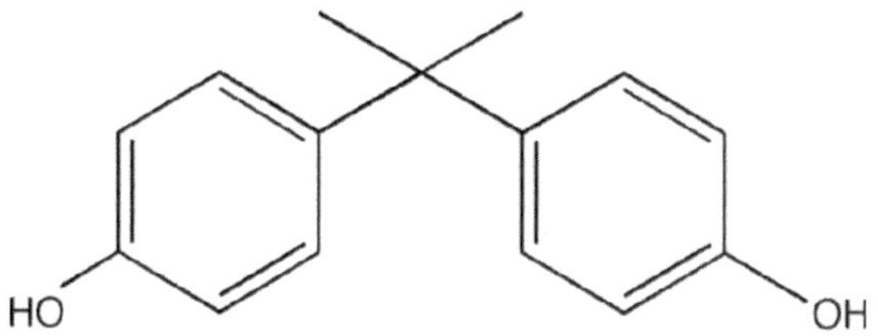

FIGURE 10.4 Molecular structure of bisphenol-A (BPA).

A group of researchers from Institut Català d'Investigació Química (ICIQ) developed a method to produce polycarbonates from citrus limonene (*cf.* Figure 10.5) and CO_2, two products that are abundant in nature. *Via* a catalytic process that allows controlled CO_2-limonene oxide coupling, it is achieved that the polymer chain presents rigid parts that produce a polycarbonate that is more sustainable and less toxic. The aim was to find alternatives to BPA, since in countries (e.g., France, Canada, Denmark) its use is prohibited because in low concentrations, it can cause imbalances in the hormonal system, etc. and it is obtained *via* non-renewable petroleum. One of the biggest advantages is that it is biodegradable, so it breaks down in a short time thanks to the actions of micro-organisms, which is an option for recycling. One of the most characteristic features of a limonene-derived biopolymer is that it presents the highest glass transition temperature ever described for a polycarbonate, making it more applicable (e.g., in the field of coatings and adhesives for packaging food, other mechanical and resistance functions).

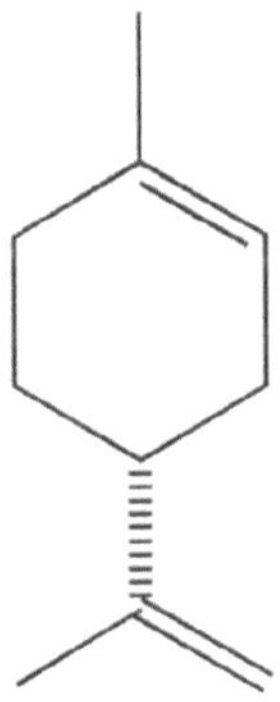

FIGURE 10.5 Molecular structure of limonene.

10.4 FINAL REMARKS

From the present results and discussion, the following final remarks can be drawn.

1. *Citrus* species belong to the agricultural biodiversity of the Valencia Region. However, the efficient mobilization of citrus biodiversity in innovative breeding schemes requires a previous understanding of *Citrus* origins and genomic structures. Findings draw an evolutionary framework for citrus, a scenario that challenges current taxonomic and phylogenetic thoughts, and points to a reformulation of the genus

Citrus. When people observe the plants that today serve them as food, they should be conscious that each one contains an evolutionary history, based on the principles of variation by mutation and selection, in part owing to its permanence as wild species and, in part, owing to the action of man, which contributed to select the genetic changes produced sometimes by reasons of searching for appropriate food, and, other times, directed by cultural or religious values.

2. Interest in the natural product-led researchers and the wellness industry to explore natural sources for alternative and complementary medicine to prevent and/or treat inflammatory diseases.
3. Chemistry establishes a bridge between the familiar and the fundamental. This report provides a general view of such a bridge but, hopefully, one can see across the fog and understand how a chemist thinks.
4. This description will teach people that the world is both simple and complex. It is complex because even a leaf is composed of an innumerable type of molecules with a variety that changes with the seasons. It is simple because scientists can unravel the leaf, distinguish its components, and see how in summer one of its components dominates and another, in autumn. The task for unraveling is far from complete, and an era can pass before knowing with certainty all the details about the differences between an orange and a banana, much less between a man and a woman. However, this account teaches people that the crowd of molecules that make up the world is less and less an amorphous worm, and increasingly a group of understandable individual personalities.

ACKNOWLEDGMENTS

The authors thanks the support from Fundacion Universidad Catolica de Valencia San Vicente Martir (Project No. 2019-217-001UCV).

KEYWORDS

- **limonene**
- **molecular distinction**
- **odor**
- **orange fruit composition**

- **taste**
- **terpene**
- **typing**

REFERENCES

1. Yu, X., Zhang, Y., Wang, D., Jiang, L., & Xu, X., (2018). Identification of three kinds of citri reticulatae pericarpium based on deoxyribonucleic acid bar-coding and high-performance liquid chromatography-diode array detection-electrospray ionization/mass spectrometry/mass spectrometry combined with chemometric analysis. *Pharmacogn. Mag., 14*, 64–69.
2. Agbor, A. M., & Naidoo, S., (2019). Plants used by African traditional healers in the management of oral diseases: A review. *Int. J. Res.-Granthaalayah, 7*(8), 273–286.
3. Wang, F., Chen, L., Chen, H., Chen, S., & Liu, Y., (2019). Analysis of flavonoid metabolites in citrus peels (*Citrus reticulata* "Dahongpao") using UPLC-ESI-MS/MS. *Molecules, 24*, 2680–1–12.
4. Powers, C. N., Satyal, P., Mayo, J. A., McFeeters, H., & McFeeters, R. L., (2019). Bigger data approach to analysis of essential oils and their antifungal activity against *Aspergillus niger*, *Candida albicans*, and *Cryptococcus neoformans*. *Molecules, 24*, 2868–1–12.
5. Olivés, J., & Mestres, J., (2019). Closing the gap between therapeutic use and mode of action in remedial herbs. *Front. Pharmacol., 10*, 1132–1–11.
6. Torrens-Zaragozá, F., (2013). Molecular categorization of yams by principal component and cluster analyses. *Nereis, 2013*(5), 41–51.
7. Torrens, F., & Castellano, G., (2014). From Asia to Mediterranean: Soya bean, Spanish legumes and commercial *soya bean* principal component, cluster and meta-analyses. *J. Nutr. Food Sci., 4*(5), 98–98.
8. Torrens, F., & Castellano, G., (2018). Principal component, cluster and meta-analyses of soya bean, Spanish legumes and commercial soya bean. In: Torrens, F., Balköse, D., & Thomas, S., (eds.), *High-Performance Materials and Engineered Chemistry* (pp. 267–294). Apple Academic–CRC Press: Waretown, NJ.
9. Torrens-Zaragozá, F., (2015). Classification of fruits proximate and mineral content: Principal component, cluster, meta-analyses. *Nereis, 2015*(7), 39–50.
10. Torrens-Zaragozá, F., (2016). Classification of food spices by proximate content: Principal component, cluster, meta-analyses. *Nereis, 2016*(8), 23–33.
11. Castellano, G., Tena, J., & Torrens, F., (2012). Classification of polyphenolic compounds by chemical structural indicators and its relation to antioxidant properties of *Posidonia oceanica* (L.) Delile. *MATCH Commun. Math. Comput. Chem., 67*, 231–250.
12. Castellano, G., González-Santander, J. L., Lara, A., & Torrens, F., (2013). Classification of flavonoid compounds by using entropy of information theory. *Phytochemistry, 93*, 182–191.
13. Castellano, G., Lara, A., & Torrens, F., (2014). Classification of stilbenoid compounds by entropy of artificial intelligence. *Phytochemistry, 97*, 62–69.

14. Castellano, G., & Torrens, F., (2015). Information entropy-based classification of triterpenoids and steroids from *Ganoderma*. *Phytochemistry, 116*, 305–313.
15. Castellano, G., & Torrens, F., (2015). Quantitative structure–antioxidant activity models of isoflavonoids: A theoretical study. *Int. J. Mol. Sci., 16*, 12891–12906.
16. Castellano, G., Redondo, L., & Torrens, F., (2017). QSAR of natural sesquiterpene lactones as inhibitors of Myb-dependent gene expression. *Curr. Top. Med. Chem., 17*, 3256–3268.
17. Torrens, F., & Castellano, G., (2020). Structure-activity relationships of cytotoxic lactones as inhibitors and mechanisms of action. *Curr. Drug Discov. Technol* (Vol. 17, pp. 166-182) in press.
18. Torrens, F., Redondo, L., & Castellano, G., (2017). Artemisinin: Tentative mechanism of action and resistance. *Pharmaceuticals, 10*, 20–4–4.
19. Torrens, F., Redondo, L., & Castellano, G., (2018). Reflections on artemisinin, proposed molecular mechanism of bioactivity and resistance. In: Haghi, A. K., Balköse, D., & Thomas, S., (eds.), *Applied Physical Chemistry with Multidisciplinary Approaches* (pp. 189–215). Apple Academic–CRC Press: Waretown, NJ.
20. Torrens, F., & Castellano, G., (2020). Chemical and biological screening approaches to phytopharmaceuticals. In: Pourhashemi, A., Deka, S. C., & Haghi, A. K., (eds.), *Research Methods and Applications in Chemical and Biological Engineering* (pp. 3–12). Apple Academic–CRC Press: Waretown, NJ.
21. Torrens, F., & Castellano, F., (2020). Chemical components from Artemisia austro-yunnanensis, anti-inflammatory effects and lactones. In: Pogliani, L., Torrens, F., & Haghi, A. K., (eds.), *Molecular Chemistry and Biomolecular Engineering: Integrating Theory and Research with Practice* (pp. 73-83). Apple Academic–CRC Press: Waretown, NJ.
22. Torrens, F., & Castellano, G., (2018). Classification of citrus: Principal components, cluster, and meta-analyses. In: Haghi, A. K., Balköse, D., & Thomas, S., (eds.), *Applied Physical Chemistry with Multidisciplinary Approaches* (pp. 217–234). Apple Academic–CRC Press: Waretown, NJ.
23. Torrens, F., & Castellano, G., (2019). Extraction of natural products found in vegetal species: Clove/*Citrus*. In: Aguilar, C. N., & Carvajal-Millan, E., (eds.), *Applied Food Science and Engineering with Industrial Applications* (pp. 305–318). Apple Academic–CRC Press: Waretown, NJ.
24. Torrens, F., & Castellano, G., (2020). World of biological activities and safety of *Citrus spp.* Essential Oils. In: Pogliani, L., Torrens, F., & Haghi, A. K., (eds.), *Molecular Chemistry and Biomolecular Engineering: Integrating Theory and Research with Practice* (pp. 113–119). Apple Academic–CRC Press: Waretown, NJ.
25. Torrens, F., & Castellano, G., (2021). Citrus: Origin, evolution, postharvest, heat, pH, filtration, and transgenic. In: Khalaf, M. N., Smirnov, M. O., Kannan, P., & Haghi, A. K., (eds.), *Environmental Technology and Engineering Techniques: Basic Concepts and Health Interventions* (pp. 155–166). Apple Academic–CRC Press: Waretown, NJ.
26. Torrens, F., & Castellano, G. (2020). *Citrus b*-Cryptoxanthyn reduces 30% body fat and nobiletin benefits. In: Talele, S. G., Shinde, R. S., Andhare, A. A., & Haghi, A. K., (eds,), *Applied Pharmaceutical Science and Microbiology: Natural Products Chemistry*. Apple Academic–CRC Press: Waretown, NJ.
27. Atkins, P., (2003). *Atkins' Molecules*; Cambridge University: Cambridge.

28. Ortolá, A., (2017). Bioplastics with limonene and without bisphenol A. *DAUALDEU, 2017*(13), 46.
29. Torrens, Z. F., (2011). Polymer bisphenol-A, the incorporation of silica nanospheres into epoxy–amine materials and polymer nanocomposites. *Nereis, 2011*(3), 17–23.
30. Torrens, F., & Castellano, G., (2013). Bisphenol, diethylstilbestrol, polycarbonate and the thermomechanical properties of epoxy–silica nanostructured composites. *J. Res. Updates Polym. Sci., 2*, 183–193.

CHAPTER 11

Innovation and Challenges in the Development of Functional and Medicinal Beverages

DAYANG NORULFAIRUZ ABANG ZAIDEL, IDA IDAYU MUHAMAD, ZANARIAH HASHIM, YANTI MASLINA MOHD. JUSOH, and ERARICAR SALLEH

Food and Biomaterial Engineering Research Group, School of Chemical and Energy Engineering, Faculty of Engineering, Universiti Teknologi Malaysia, 81310 Johor Bahru, Malaysia, Tel: +607 5535503; E-mail: dnorulfairuz@utm.my

ABSTRACT

Recent reports have recognized that functional foods are the most emerging interest in food sectors. In particular, the functional beverage market is one of the fastest-growing segments within this sector. Beverages are categorized as the most dynamic functional food due to their convenience and the high possibility to meet market needs. Functional beverages have great potential to be incorporated with desirable nutritional ingredients and bioactive compounds such as vitamins, minerals, antioxidants, fatty acids, fiber, prebiotics, and probiotics. This chapter highlighted the recent innovation and development of functional and medicinal beverages including the current processes and methods of developing successful functional and medicinal beverages. The topics include the definition of functional and medicinal beverages and the type of functional beverages available in the market, the market trends and overview, emerging trends in processing technology, quality, and safety issues in functional beverages prior to commercialization of products. It has become a challenge to maintain a good quality of functional and medicinal beverages especially in terms of its stability, shelf life, functionality, and safety. These issues must be taken seriously for improving the quality and health benefits of the beverages, as well as for the future of the food industry.

11.1 INTRODUCTION

Recent reports have recognized that functional foods are the most emerging interest in food sectors with the functional beverage market, particularly being the fastest-growing segment within this sector. Nowadays, as people are more health-conscious, the invention of foods and beverages is focused more on their function towards the prevention and treatment of disease has been growing rapidly. The consumption of functional foods and beverages has been one of the top choices for many people as they target a number of health objectives beyond the basic nutritional functions.

Functional beverages are promoted with many health benefits such as improved immunity and digestion systems, energy-boosting, weight management, improving mental focus, and preventing pain that is associated with bone and joint conditions. In addition, some beverages are formulated specifically for age demographics and genders, such as products targeted on kids, women, and seniors. This innovation and diversification in beverages have improved the beverages not only in terms of taste but also in nutritional value and functionality.

Beverages have been categorized as the most dynamic functional food as it provides convenience and opportunity in reaching the consumer demands. The composition, physical attributes, the dimensions of size and shape, and the easiness of delivery and storage at either ambient or refrigerated temperature are among the acceptable preferences of beverage products. Besides that, it has great potential to be incorporated with desirable nutritional ingredients and bioactive compounds such as vitamins, minerals, antioxidants, fatty acids, plant extracts and fiber, prebiotics, and probiotics. Functional beverages are a brilliant medium for conveying nutritional ingredients and bioactive compounds to consumers in a natural way.

11.2 DEFINITION OF FUNCTIONAL AND MEDICINAL BEVERAGES

The term "functional food" has no legislative definition in most countries. Mark-Herbert [110] and Niva [120] observed that because of the lack of a legitimate definition, food and nutrition experts find it challenging to distinguish between foods that are functional and those that are conventional. Up until now, academic bodies, national authorities, and even the food industry have defined functional foods differently—some with very simple definitions while others preferring more complex ones. For example, functional food has been simply defined as "foods that provide not only basic nutrition but also other health benefits," while a complex definition of functional food is "food that resembles conventional food meant for normal diet consumption, albeit modified to provide more than simple nutrient requirements to serve other physiological roles" [15].

In the mid-1980s, Japan introduced the first concept of functional foods, namely foods that had ingredients specifically for health functions or health use, also known as FOSHU [98]. According to Bigliardi and Galati [18], the specific definition of FOSHU, as given by the Japanese Ministry of Health, Labor, and Welfare, is "any food with a particular label and license that promises specific benefits to the health of the persons that consume it for health uses." Diplock et al. [46] further defined functional foods and drinks as foods that "provide beneficial effects for bodily functions in addition to basic nutrition such that the health of the person consuming it will be improved and the risk of disease is reduced."

In 1998, a report by the Functional Food Science Commission in Europe defined "foods that sufficiently prove to benefit any functions of the body in addition to providing nutritional benefits such that health is improved and the risk of disease is reduced" as functional foods (including drinks). Meanwhile, Pravst [136] noted the combined definitions of functional foods provided by other studies such as "food that is natural and unmodified; enhanced food with components that have been improved via biotechnological means, special breeding, or unique growing conditions; food that has added beneficial components; food that provides unique benefits as a result of the removal of components via technological or biotechnological means; food with favorable properties as a result of component replacement with another better component; food with a beneficial component modified via technological, chemical, or enzymatic means; food with a component that was modified to become biologically available; or any food with the above combinations."

In addition to the above definition, the International Life Science Institute (ILSI) Europe through the European Commission's Concerted Action on Functional Food Science in Europe (FuFoSE) defined functional food as "food products that, in addition to providing basic nutrition, also benefit any human bodily function to bring about improved condition, whether general of physical, and causing the risk of disease to reduce."

Diplock et al. [46] further emphasized that functional food should be consumed as per other food that is naturally consumed as part of one's diet. In other words, functional food is not taken as a capsule or pill but rather in its natural (food) form. Nevertheless, Ohama et al. [123] disagreed with this definition, stating that Japanese FOSHU products had been consumed as tablets or pills since 2001, while other functional foods still retain their normal form. Elsewhere, Yang [177] contended that China defines functional food as "foods that provide special health benefits or foods that provide minerals or vitamins to the body, regulates the functions of the human body, and are consumed by specific groups of people, but are ultimately not meant for therapeutic purposes, and as a result of consumption, will not cause any acute, sub-acute, or chronic harm." Canada, on the other hand, defines "food that resembles (or may even be) conventional food that is part of a regular diet consumption that proves to benefit humans physiologically and decreases chronic disease risks in addition to providing basic nutrition" as functional food [98].

In the US, however, functional food is regulated by the Food and Drug Administration but is not defined legally. According to the Academy of Nutrition and Dietetics in the US, functional foods are "foods that are enhanced, fortified or enriched, or whole foods, that have shown significant evidence of providing effective benefits to health when taken as part of a regular varied diet." Based on the above definitions, functional foods or beverages can be simply defined as "food products that have additional physiological benefits over and above their benefits that occur naturally."

11.3 NUTRITIONAL AND HEALTH CLAIMS

From a regulatory point of view, functional foods do not have an appropriate definition and category. Brown and Chan [24] reported that any regulatory authority with the exception of Japan has classified "functional food" as a distinguishable product. Therefore, Sun-Waterhouse [159] revealed that there is an uncertainty in labeling the new product, either as food, supplement, or drug. It is important that before a new functional food product to enter the market, clear descriptions of functionality or health and nutritional effect of this food should meet the fundamental specifications.

In Europe, it has been stated that "nutrition claim" is made to recommend the specific valuable nutritional composition of food including "source of," "free of," "reduced," while "health claim" is documented to propose the synergistic effect within health and food group, a food, or one of its compounds. Dolan [47] concise that the claims are defining "generic-health claim" is associated to the important role of nutrient and other ingredients for growth, development or function of the body; psychological and behavioral performances; weight management, fullness or lessening of existing energy and "health claim" which described the ability of the food in lowering the chance of having disease and children's growth failure.

In the U.S., functional food products must follow the guidelines for conventional foods. According to Jackson and Paliyath [79], it is a requirement to ensure that the product is safe to be advertised as foods, and the incorporated ingredients are "generally recognized as safe," GRAS, or permitted as food additives. In Hasler [71], functional food product claims can be classified into three; "health claims," "nutrient content claims," and "structure or function claims."

The registered products in the U.S. which are a claim for dietary supplement ingredients are applicable to be used worldwide. However, the claims of certain U.S. supplement ingredients have not been approved in Europe and vice versa because of the distinct regulations applied. Dolan [47] reported that the approval of the products needs immense processes involving the national authorities, EFSA, and the European Commission which are the main restrictions to advertise the product with claims in Europe.

11.4 DEVELOPMENT OF FUNCTIONAL AND MEDICINAL BEVERAGES

Development of a functional beverage includes the extraction of functional ingredients or active compounds, formulation of the beverage, processing method, or technology used and quality assessment of the beverage prior to commercialization. Each step in producing functional or medicinal beverages is important to make sure that the quality and safety of the beverages are the top priority. The beverage quality includes shelf-life, stability, storage, and packaging which may then contribute to the health of consumers. Different types of functional and medicinal beverages vary from dairy-based, vegetable and fruits drinks, energy drinks, sport and performance drinks, soy-based, and herbal drinks. These different types of beverages are formulated based on different functional ingredients and can be produced using different processing methods. As the modern technology has evolved, more innovation on functional beverages has been reported to improve the processing and preserve the functional ingredient or active compound in the beverages. Many studies have been reported on the extraction of functional ingredients or active compounds from natural sources either from animals or plants for application in beverage production [43, 127, 137].

11.5 TYPES OF FUNCTIONAL AND MEDICINAL BEVERAGES

Functional and medicinal beverages appeared to be the most active functional foods category due to convenience, health benefits, ease of distribution, and storage, as well as ease of fortification of products with desirable nutrients and bioactive compounds [128]. This chapter highlights several types of popular beverages available in this fast-growing sector, namely dairy-based beverages, vegetables and fruit beverages, energy drinks, sport and performance drinks, soy-based beverages, herbal drinks, and ready-to-drink tea.

11.5.1 DAIRY-BASED BEVERAGES

Yogurt drinks, fresh milk, and fermented milk are the example of dairy-based beverage products that are often regarded as an excellent vehicle for probiotics. According to FAO/WHO [189], probiotics are living microorganisms which when consumed in a specific dose will provide beneficial health effects for the host. In fact, the individual probiotics give various health effects including immune response enhancement, cholesterol level reduction, colon cancer prevention, irritable bowel syndrome (IBS) alleviation, inflammable bowel disease (IBD) alleviation, and antibiotic-associated diarrhea treatment, among others [129].

Today, with regard to the aforementioned health benefits, many dairy manufacturers began to incorporate probiotic bacteria (i.e., *Lactobacillus* spp., *Bifidobacterium* spp., etc.) into their products [41]. The main reason behind such action is to satisfy consumers' demand at the same time to gain a competitive advantage in the world market. Besides probiotic bacteria, some producers may fortified/enriched dairy beverages with bioactive components (i.e., Omega-3, phytosterols, fiber, minerals, vitamins, etc.) in order to enhance the products' functional properties [22]. Table 11.1 lists some examples of functional dairy-based beverages available in the market.

TABLE 11.1 Examples of Commercially Available Functional Dairy-Based Beverages [41, 129, 135]

Brand	Producer	Active compounds
PROBIOTICS		
Yakult®	Yakult Honsha Co. (Japan)	*L. casei* Shirota
Vitagen®	Malaysia Milk Sdn. Bhd. (Malaysia)	*L. acidophilus* *L. casei*
Chamyto®	Nestle (France)	*L. johnsonii* *L. helveticus*
ENRICHED BEVERAGES		
Evolus®	Valio Ltd. (Finland)	Bioactive peptides
Heart Plus®	PB Food (Australia)	Omega-3
Benecol®	Mc Neil Nutritionals (United Kingdom)	Phytosterol
Dairyland Milk-2-Go®	Saputo (Canada)	Omega-3 Calcium Vitamins
Night-Time Milk®	Cricketer Farm (United Kingdom)	Melatonin
Zen®	Danone (Belgium)	Magnesium
Natural Linea®	Corporacion Alimentaria Penanata S.A. (Spain)	Conjugated linoleic acid

Functional dairy-based beverages have continued to be as essential as the primary food in the worldwide, particularly in the Western parts. However, several health risks are associated with these products. According to Kumar et al. [96], the major drawback of dairy beverage intake appears to be lactose intolerance. The condition when lactose, a milk sugar contains abundantly in dairy products, is unable to digest because of the lack of the lactase enzyme in the small intestine is called lactose intolerance [44]. People with lactose intolerance when consuming dairy products will develop symptoms like flatulence, bloating, loose stool, and cramping [164]. In this case, the market size for dairy companies will be reduced. Thus, the companies are launching more alternatives that could provide the same nutrients and health benefits as dairy products so that the range of their targeted customers will become larger.

11.6 VEGETABLES AND FRUIT BEVERAGES

Dairy-based beverages have long been considered as an ideal delivery vehicle of nutrition for humans. However, lactose intolerance appears to be the major drawback of functional dairy products. According to Lomer et al. [105], lactase non-persistence is suffered by up to 70% of the total population of the world which may negatively affect the overall market of those products.

Unlike dairy-based products, vegetable and fruit-based functional beverages are regarded as safe and healthy drinks by consumers particularly to those who are sensitive to dairy products. Additionally, the vital nutrients of vegetables and fruit juices, such as vitamins, minerals, phytochemicals, antioxidants, etc. could provide an excellent medium for probiotic [67]. The global market for fortified vegetables and fruit juices is expected to reach 65.9 billion liters by 2022, driven by the growing awareness over the health benefits of juicing fruits and vegetables [192]. Blueberry, strawberries, cranberry, grapes, blackcurrant, cherries, pomegranate, mango, apple, acai, guarana, kiwifruits, peach, and plums are examples of fruits that are often utilized in industrial productions [159]. Based on the reports from Gaanappriya et al. [60], orange and watermelon juices were proven to be a suitable carrier of lactobacilli (for the purpose of functional beverage preparation) for individuals who are sensitive of dairy products. Table 11.2 shows several instances of vegetable and fruit-based functional beverages that are accessible in the food industry which are fortified with probiotics and enriched with vitamins and minerals.

Apart from nutrients that contribute to the health of mankind, most commercialized vegetables, and fruit juices contain a high amount of sugar for preservation purposes as well as to enhance the taste of products [96]. Nevertheless, this raised the health concern of consumers as long-term consumption of high-calorie products may increase the risk of diabetes and heart-related complications. Thus, manufacturers nowadays are attempting to adapt modified or other new approaches in order to develop a novel, low calorie fortified vegetable or fruit juice containing high nutritional value at the same time maintaining the taste that able to satisfy consumers' preferences.

TABLE 11.2 Examples of Commercially Available Vegetable and Fruit-Based Functional Beverages [41, 129]

Brand	Producer	Active compounds
PROBIOTICS		
Whole Grain Probiotic Liquid®	Grainfields (Australia)	Vitamins Amino acids Enzymes Probiotics: - *L. acidophilus* - *L. delbrueckii* - *Saccharomyces cerevisiae* var. *boulardii* - *S. cerevisiae*
Biola®	Tine BA (Norway)	Probiotics: *L. rhamnosus* GG
Gefilus®	Valio Ltd. (Finland)	Vitamins C and D Probiotics:
-		*L. rhamnosus* GG
ENRICHED BEVERAGES		
Tomato Juice Plus®	Langer Juice Co., Inc. (U.S.A.)	Vitamins Minerals
Tropicana Essentials Orange Juice & Calcium®	Tropicana (U.S.A.)	Calcium
Welch's 100% Grape Juice with Calcium®	Welch Foods Inc. (U.S.A.)	Calcium

11.7 ENERGY DRINKS

Beverages loaded with stimulant drugs such as caffeine are well known as an energy drink and often targeted to young people aged within 21 and 35 years to provide mental and physical stimulation [41]. According to Duncan and Hankey [48], more than 40% of athletes consume energy drinks to boost their athletic performance. Besides caffeine, vitamin B complex and taurine are the most common ingredients for such a product. Taurine, can be found naturally in breast milk, meat and fish is an amino acid that can be obtained naturally in the body. Several investigations found that supplementation with taurine could enhance athletic performance whereas when mixed with caffeine could enhance mental performance [140]. Examples of commercialized energy drinks are Red Bull® (Red Bull GmbH, Austria), Monster Energy® (Hansen Natural Corp., U.S.A.), and Full Throttle® (Coca-Cola Co., U.S.A.) [41].

Some energy drinks may include herbal extracts as an ingredient to enhance the effect of their products. Examples of herbal extract usually used are ginseng, ginkgo biloba, guarana, etc. Ginseng is a type of herb that has been used for over 2,000 years in China, Korea, and Japan, as a remedy for promoting longevity as well as enhancing physical and mental conditions [88]. In addition, ginseng can be used as an anti-stress and antioxidant agent. However, the effect might be depending on the dose and frequency of consumption since this herb has multiple drug interactions [70].

It has been reported that energy drinks might also be associated with health risks, for an instance the raised risk of alcohol-related harm, and immoderate drinking may lead to psychiatric and cardiac conditions [151]. Adults physical performance has been revealed to be a boost when taking caffeine by improving the magnitude of response to aerobic resistance and strength, slow down exhaustion, and improving reaction time [30]. Nevertheless, these effects are extremely inconsistent and depend on the intensity of caffeine whereas their effects are not being investigated on children and adolescents [39].

11.8 SPORTS AND PERFORMANCE DRINKS

Sports drinks are flavored beverages that are particularly formulated to help consumers/athletes to replenish water, electrolytes (minerals such as potassium, sodium, magnesium, chloride, and calcium), and energy, particularly after exercise. Unlike energy drinks, sport and performance drinks do not contain caffeine [39]. Generally, sports and performance drinks can be categorized into three major types, namely, isotonic, hypertonic, and hypotonic, all of which comprise of different amounts of fluid, electrolytes and carbohydrates, and different functions as shown in Table 11.3.

Glucose, sucrose, fructose and maltodextrin/glucose polymers, or a combination of these appear to be the common ingredients for sports and performance drinks. According to Campbell [27], the advantage of adding maltodextrin/glucose polymers is that the lower sweetness of these compounds increases the concentration of carbohydrates without increasing the sweetness of the product. Nevertheless, when high carbohydrate-containing drinks are taking in immoderate, it may increase an individual's intake of total daily caloric which will eventually gain extra weight, dental caries, and low quality of diet. Table 11.4 illustrated several examples of commercially available sport and performance drinks.

TABLE 11.3 Difference Between Three Major Types of Sports and Performance Drinks Available in the Market [184]

Types of sports and performance drinks	Isotonic	Hypertonic	Hypotonic
Level of fluid, electrolytes and carbohydrate	Same concentrations of salt and sugar as in the human body	Higher concentration of salt and sugar than the human body	Lower concentration of salt and sugar than the human body
Function	To replace fluids lost through sweating quickly.	To supplement daily carbohydrate intake.	To replace fluids lost through sweating
	To supply a boost of carbohydrate	Top-up muscle glycogen stores	
Recommended to...	Most athletes, including middle and long-distance running.	Individuals involved in ultra-distance events, but must be used in conjunction with isotonic drinks to replace lost fluids	Athletes who require fluid without a carbohydrate boost, such as gymnasts
	Individuals involve in team sports		

11.9 SOY-BASED BEVERAGES

Soy-based beverages are aqueous extracts of whole soybeans (Glicine max) and a significant source of high-quality protein. They are frequently used as a dairy substitute, especially individuals who are allergic to milk or having severe intolerance to lactose because soy-based beverages contain almost similar nutrients and health benefits as in milk and dairy products. Although soy extracts lack calcium, the present soy-based beverages have this problem solved by adding artificial calcium to them where this will further increase their nutritional values [67]. Besides that, these plant-based products are also a suitable alternative to be offered to vegetarians. Soy-based beverages have wide applicability in the food industry, such as they can be used directly as a beverage or as an ingredient in desserts, infant formulas, and yogurt [34].

TABLE 11.4 Examples of Commercially Available Sport and Performance Drinks [41, 72]

Brand	Producer	Active compounds
Powerade®	Coca-Cola Co., (U.S.A.)	Sodium, Iron
Powerade Zero®	Coca-Cola Co., (U.S.A.)	Sodium, Potassium, B-vitamins
Accelerade®	Pacific Health Laboratories Inc., (U.S.A.)	Sodium, Potassium, Calcium, Vitamin E, Protein
All Sport Body Quencher®	All Sport. Inc., (U.S.A.)	Sodium, Potassium, Vitamin C
Gatorade Endurance®	PepsiCo Inc., (U.S.A.)	Sodium, Potassium Calcium, Magnesium

Products of this category have become increasingly popular among consumers ever since the development of new technologies that can reduce the usual astringent taste in soy products [37]. According to the statistics provided by Statisca [194], the global sales volume of soy-based beverages amounted to 13.48 billion liters in 2015 and this was expected to increase to 16.29 billion liters in 2018. Thus, many food manufacturers are constantly launching new products in the soy beverage category nowadays, thereby widening the product varieties available for consumers in the market. Some of the key players of the soy food products market are Whole Soy & Co., Northern Soy Inc., DuPont Solae, ADM Inc. Solbar Ltd., The Scoular Company, Cargill Inc., and Linyi Shansong Biological Products Co. Ltd. [67].

Soy-based beverages have relatively fewer calories and various health benefits. Consumption of these products can reduce the risk of heart failure as soy extracts reduce the cholesterol level in the blood [148]. Other potential health benefits recently discovered in various studies are related to osteoporosis, menopausal symptoms, breast cancer, and prostate cancer [125, 165]. In addition, soy-based beverages also provide minerals (i.e., zinc and iron) and vitamins (i.e., B6 and B12). Vitamin B6 and B12 are essential for increasing hair growth as they nourish the hair follicles [143]. Nevertheless, individuals with kidney stones containing oxalates should avoid consuming these beverages as they are a rich source of oxalates [190].

11.10 HERBAL DRINKS

Herbal drinks are beverages made from the infusion or decoction of medicinal plants/herbs in hot water. Various parts of the plant can be used to prepare such beverages; for instance, roots, stems, bark, leaves, fruits, or flowers. One popular product currently available in the market that falls under this category is herbal tea.

Herbal teas, also called tisanes are the second largest consumed beverage in the world after water. Some of the major players in the herbal tea market are Nestea, Global Herbitech, Tata Global Beverages, Green Earth Products Pvt Ltd., AB Food and Beverages, and Buddha's Herbs [186]. Herbal teas are free of caffeine. They are often made from one main or a combination of several herbal ingredients, designed to bring about a specific therapeutic and energizing effect, such as promoting a good night's sleep, relieving stress, supplying antioxidants to the body, supporting heart health, helping to prevent colds, strengthening the immune system, providing cleansing properties for the body, etc. [138]. Table 11.5 shows various types of herbal teas and their specific health benefits.

TABLE 11.5 Different Types of Herbal Teas and Their Specific Health Benefits [35, 138, 142, 156]

Types of Herbal Tea	Health Benefits
Ginger tea	Stimulates and soothes the digestive system.
	Contains anti-inflammatory properties, which are helpful for arthritic people.
	Aids people experiencing nausea of any kind, such as motion sickness, morning sickness, etc.
Peppermint tea	Treats upset stomach effectively when combined with chamomile.
Chamomile	Promotes a good night's sleep.
	Soothes stomach pains and acts as a gentle laxative.
	Alleviates menstrual cramps.
Cardamom tea	Relieves coughs.
	Provides a calming effect to women who experience mood swings during their menstrual period.
	Treats indigestion, relieves flatulence, and prevents stomach pain.
	Reduces nauseous feeling.
Green tea	Provides anti-aging benefits.
	Prevents certain types of cancers, such as breast, prostate, stomach, skin, throat, and lung cancer.
	Aids in weight loss.
	Fights tooth decay and gum diseases.
	Reduces the LDL (bad) cholesterol level in the body.

While these products offer many beneficial health effects, there are several risks associated with the consumption of herbal teas. Studies revealed that sassafras tea contains 80% of the toxic compound known as safrole where a large amount of safrole intake may lead to liver damage [50]. Additionally, Rodriguez-Fragoso et al. [142] have found that long-term consumption of peppermint tea might lower the testosterone level in adult males, thereby affecting their fertility. Ernst [51] has proven that pyrrolizidine alkaloids detected in coltsfoot tea (*Tussilago farfara*) are toxic, and consumption of this tea may increase the risk of liver toxicity and a person's chance of developing cancers.

Nevertheless, the safety of many herbal drinks remains unknown, as toxicological studies on herbal teas are very limited. More studies on the toxicity and herb-drug interactions of medicinal should be performed. Thus, it is advisable that widespread consumption of herbal infusions should be minimized until data on the levels and varieties of toxicant carcinogens, and mutagens are made available.

11.11 READY-TO-DRINK TEA

Ready-to-drink (RTD) tea refers to tea-based beverages that are sold in a prepared form and are ready for consumption. It is one of the world's fastest-growing soft drinks categories, driven by the rising consumers' demand for healthy, low sugar beverages. According to Fortitech [191], the fortified RTD tea market has grown from $1.4 billion to $2.6 billion from 2004 to 2009. The market continues to increase steadily ever since then. In 2016, RTD tea appeared to be one of the most popular beverages in Malaysia where an estimated of 15% off-trade volume growth was achieved [187]. Additionally, in the same year, RTD tea performed strongly in Australia as the retail volume sales increased up to 7% [188]. The key vendors in the global RTD tea market include Sunny Delight Beverages Co. (SDBC), Unilever, Talking Rain, and TeaZazz, while Fraser & Neave Holdings Bhd. dominates the RTD tea market in Malaysia [187, 188].

RTD teas, regardless of the types, are convenient and most offer a healthy alternative to other ready-to-drink beverages on the market. Consumption of these products may aid in a person's quest to manage a myriad of health conditions ranging from weight loss to cognitive function and heart health. Besides that, RTD teas are desirable vehicles for fortification where even more health benefits can be supplemented to this timeless drink.

Matcha, one of the latest consumer trends in RTD tea products, is a powder ground from raw green leaves called tencha and is made into a nutrient-packed form of green tea. Matcha has been promoted to contain polyphenol, particularly flavonols and flavanols, which represent 30% of fresh leaf dry weight [112, 116]. Some of the possible health benefits gained from matcha intake are the prevention of cancer and cardiovascular diseases, anti-oxidative, anti-inflammatory, anti-bacterial, anti-arthritic, etc. [161]. Other examples of RTD teas are jasmine hibiscus RTD tea, passion fruit iced green tea, and green tea with cactus water, among others [12].

11.12 FUNCTIONAL INGREDIENTS IN BEVERAGES

Functional ingredients are defined as ingredients that can provide health benefits for increased well-being. The micronutrients including antioxidants, probiotics, prebiotics, vitamins, and minerals that are included in a healthy beverage are reported to have physiological benefits and ability to reduce the risk of chronic diseases more advanced than a macronutrients function [139]. Regularly, functional ingredients are loaded in a beverage with a small amount of less healthy ingredients such as sugar and fats to produce functional beverages. This is done in order to obtain health benefits and at the same time to develop, a stable beverage that has appealing flavors throughout the shelf-life [54]. The functional properties and benefits of these ingredients can influence the consumers' awareness when they are purchasing the functional beverages. Each consumer had their own different reasons on why they are consuming the beverages and these because of either to treat and prevent certain diseases or just to absorb nutrients for keeping and maintaining a healthy eating pattern [20].

11.12.1 ANTIOXIDANT (A, C, E, ZINC)

Antioxidants are molecules that are able to prevent the oxidation of other molecules. There are increasing consumer demands for food and beverage containing high antioxidants in the food and beverage sector. Research shows that food-containing antioxidants are associated with better long-term health advantages including the improvement of immunity performance and the reduced risk of cancer and heart disease. Drinking high antioxidant beverages may help to protect against aging, Alzheimer's disease, and other chronic diseases [97, 118]. Vitamins A, C, E, and zinc are a good source of antioxidants [97].

Vitamin A is a type of fat-soluble vitamin and it plays an important role in maintaining healthy vision and skin, neurological functionality, and reducing inflammation through fighting free radical damage. Nutritionally, vitamin A required by the human body can be obtained from two primary sources which are from animal sources and plant sources [33]. The animal-derived sources can be obtained in a form of retinol (active Vitamin A), which can be used directly by the body. For plant-derived Vitamin A, which can be obtained from fruits and vegetables, it is in the form of pro-Vitamin A such as beta-carotene and related carotenoids, which have to be converted to retinol after being consumed in order to be fully utilized by the body [127].

Vitamin C or L-ascorbate, a water-soluble antioxidant that is widely distributed in nature. It is highly abundant in some fruits and vegetables and is also found in animal organs such as the liver and kidney [55, 127]. Usually, fruits and vegetables serve as the main sources of Vitamin C for the human body. Farbstein et al., [55] had explained that Vitamin C is needed for many physiological processes and has shown a great potential in improving immunity performances, as well as lowering the risk of cardiovascular disease and cancer. Vitamin C plays an important role in human health as an antioxidant which can protect the body against oxidative stress [122].

Vitamin E is a fat-soluble vitamin and consists of eight groups of lipophilic antioxidant compounds, which are comprised of four tocopherols compound and another four tocotrienols compound. The vitamin was easily obtained from green vegetables, grains, nuts, and various vegetable oils such as wheat germ, sunflower seed, and maize oil. Vitamin E can also be obtained in eggs and milk. Tocopherols and tocotrienols expressed a variety of biological properties such as antioxidant activity, the potential to modulate protein function and gene expression, helps in forming blood cells and boost the immune system [33, 55, 127].

11.12.2 NUTRACEUTICAL INGREDIENTS

Nutraceuticals terms are used to define health-promoting foods or components that are extracted from them. The term "nutraceutical" was first derived from "nutrition" and "pharmaceutical," and defined as "a food or part of a food that can provide medical or health benefits, including the prevention and/or treatment of a disease" [33, 97]. Nutraceutical products may be available in the form of individual, isolated nutrients, herbal products, and dietary supplements. Examples of nutraceutical ingredients are dietary fibers, polyunsaturated fatty acids, and polyphenols such as carotenoids [33].

Dietary fiber is the part of plant material that is unsusceptible to the enzymatic digestion in the digestive tract but is able to be digested by microflora in the gut. Non-starch polysaccharides (NSP) is a common source of dietary fiber, for example, celluloses, lignin, hemicelluloses, gums and pectins, resistant dextrins and resistant starches. Foods that are rich in soluble fibers include fruits, oats, barley, and beans [43, 45].

Polyunsaturated fatty acids are crucial to the body to function effectively and these are introduced externally through the diet. PUFAs can be divided into two classes of essential fatty acids which are omega-3 fatty acids and omega-6 fatty acids. PUFAs can be found in fishes such as mackerel and salmon, flaxseed, soybeans, canola, vegetable oils such as sunflower and corn oil, meat, poultry, and eggs [43]. PUFAs especially omega-3 fatty acids provide benefits in many areas of health including premature infant health, bipolar and depressive disorder, and inflammatory diseases such as asthma [69]. PUFAs are also being associated with cardiovascular protection and able to reduce morbidity and mortality from cardiovascular diseases due to its anti-inflammatory and anti-atherogenic properties [111].

Phytochemicals are available in plant and contain many types of polyphenols whereas will be detected as secondary metabolites and also must be protected from photosynthetic stress. Polyphenols can be distinguished according to approximately 8,000 groups and the well-known polyphenols such as flavonols, flavones, flavan-3-ols, flavanones, and anthocyanins had been widely investigated. Polyphenols could promote many functional properties including antioxidant, anti-inflammatory, anti-microbial, cardioprotective activities, as well as a protective effect on neurodegenerative diseases [43]. Some of the medicinal plants such as feverfew (*Tanacetum parthenium*) has been used as a source of nutraceuticals in the production of a functional beverage with anti-inflammatory properties (Carbo et al., 2014). Generally, nutraceutical able to offer physiological benefits and provide protection against chronic diseases [97].

11.12.3 VITAMINS (B, D, K)

Vitamin can be described as a group of organic, heterogeneous nutrients that are required by the human body in a very small amount [127]. Vitamin can be categorized into two groups which are either fat-soluble vitamins or water-soluble vitamins. Other than Vitamin A, C, and E which can provide antioxidant benefits, another group of vitamins such as Vitamin B, D, and K also can provide health benefits to the human body. Different types of vitamins and the sources are presented in Table 11.6.

TABLE 11.6 Commonly Used Synonyms and Example of Sources of Vitamin B, D, and K [33, 127]

Vitamin	Synonyms	Example of sources
Fat-soluble vitamins		
Vitamin D_2	Ergocalciferol	Fish liver oil, wheat germ oil, egg yolk, milk, butter
Vitamin D_3	Cholecalciferol	
Vitamin K_1	Phylloquinone, phytomenadione	Cabbage, cauliflower, tomatoes
Vitamin K_2	Menaquinone, farnoquinone	
Vitamin K_3	Menadione	
Water-soluble vitamins		
Vitamin B_1	Thiamin	Cereals, pulses
Vitamin B_2	Riboflavin	Nuts, yeast
Vitamin B_3	Pantothenic	Liver, meat, yeast
Vitamin B_5	Nicotinic acid	Yeast, egg, milk
Vitamin B_6	Pyridoxal, pyridoxine, pyridoxamine	Red meat, liver, cod roe and liver, milk and green vegetables

11.12.3.1 VITAMIN B

An emerging interest toward Vitamin B is observed as it is advertised as a source of nutrients for energy which usually found in energy drinks [97]. One of the examples of Vitamin B is Vitamin B1 which is also known as thiamine and widely distributed in living tissue. Mostly, it could be found in the form of a phosphorylated group in animal products and it is identified as a non-phosphorylated form in plant products [127]. Vitamin B2 or riboflavin are widely distributed in all plant and animal cells. This type of vitamin occurs naturally in two types of form, which are riboflavin mononucleotide and flavin adenine dinucleotide. Plants and many bacteria can synthesize riboflavin and it is available as dietary amounts in dairy products [127].

Another type of vitamin B is Vitamin B6. This vitamin's activity is shown by pyridoxine which consists of three different compounds that are pyridoxol, pyridoxal, and pyridoxamine. Generally, Vitamin B is essential energy production by helping in converting food into energy and helps in maintaining the brain function. Vitamin B is important in neurologic and for various nervous system functions, and it also helps to produce an essential protein [33].

11.12.3.2 VITAMIN D

Vitamin D is abundant consists of dairy foods and beverages. It presents in nature in several forms, which are Vitamin D2 and Vitamin D3. The dietary vitamin D occurs predominantly in animal products with a very small amount gained from plant sources. Vitamin D2 or ergocalciferol is produced by the ultraviolet irradiation of ergosterol. This vitamin is widely distributed in plants and fungi. Vitamin D3, also known as cholecalciferol, is derived from ultraviolet irradiation of 7-dehydrocholesterol and can be found in the skin of animals, including humans. Human requirements for this vitamin are obtained both from the endogenous production in the skin and from dietary sources [127]. Vitamin D is essential for the formation of teeth and bones. It helps the body to absorb and use calcium in an effective manner [33].

11.12.3.3 VITAMIN K

Vitamin K occurs in a number of forms. Vitamin K1 (phytomenadione or phylloquinone) is found in green plants and vegetables, potatoes, and fruits, while vitamin K2 (menaquinone) can be found in animal and microbial materials. Vitamin K is essential in blood clotting and bone metabolism [33].

11.13 BOTANICALS INGREDIENTS

Botanicals can be defined as fresh or dried plants, plant parts, or collective chemicals isolated from plants, which often obtained in the form of concentrated extracts. They are commonly added and used in the food and beverage manufacturing for flavor, fragrance, or other technical properties, such as coloring, thickening, or preservative [181]. Examples of botanical extracts are green tea and aloe vera extracts.

11.13.1 GREEN TEA EXTRACT

Tea is one of the most popular beverages that are consumed by most of the people around the world. Tea that is produced from plant *Camellia sinensis*, is consumed daily as green, black, or Oolong tea. Among these three types of tea, green tea has been observed to provide the most significant effects on the human health [32]. Green tea comprises of polyphenols and four catechins which are epigallocatechin (EGC), gallocatechin (GC), epicatechin (EC), and epigallocatechin gallate (EGCG). Catechins contain a huge amount of antioxidant properties and is also recognized as tea flavonoids [97]. These catechins are reported to provide various health impacts that involve decreasing the risk of Alzheimer's disease, to prevent many types of cancers including stomach, lung, pancreas, and kidney cancer, and also can act as anti-tumorigenic agents. Other than that, the consumption of green tea catechins can also prevent cardiovascular and oral diseases and able to enhance the metabolism rate. The popularity of green tea extract in non-alcoholic beverages other than tea had been increasing due to it may supply a natural source of caffeine [32, 97].

11.13.2 ALOE VERA

Aloe vera (*Aloe barbadensis*) has long been used in health foods and beverages as well as for medical and cosmetic purposes. It has been widely used due to its aromatic properties, bitter taste, and other pharmacological activities such as reduction of inflammation and wound healing [95, 137]. Besides providing medicinal values, Aloe vera also contains various nutrients such as vitamins, amino acids, minerals, enzymes, sugars, saponin, and fatty acids. These nutrients are able to provide positive effects on human health and can be used in food formulation as a functional ingredient for health benefits. Examples of Aloe vera beverage products are aloe soft drink, diet drink with soluble fiber, hangover drink, tropical fruit juice with Aloe vera, and yogurt drinks. Other than that, Aloe vera can be turned into Aloe vera desserts with chunks of aloe, instant Aloe vera granules, Aloe vera gums for sore or bleeding gums, Aloe vera candy and Aloe vera smoothies [137].

11.14 PREBIOTICS AND PROBIOTICS

Prebiotic and probiotic can be used to further improve the functionality and quality of the functional beverage. A non-digestible food ingredient is called prebiotics which was beneficially affecting the host by selectively stimulating the growth or activities of useful intestinal bacteria in the colon, by providing growth enhancers and nutrients to probiotic bacteria. Basically, prebiotic is mainly obtained by extraction from plants, followed by enzymatic hydrolysis. Compounds, such as lactulose and lactitol, fructooligosaccharides (FOS), inulin, galacto-oligosaccharides, and oligosaccharides from soy and levans are grouped as prebiotics [41, 169].

Probiotics are described as "a microbial preparation which contains live and/or dead cells including their metabolites which is intended to improve the microbial or enzymatic balance at mucosal surfaces or to simulate immune mechanisms" [97]. Probiotic cultures are beneficial for human health because it helps to the improvement of intestinal microbial balance which results in the inhibition of bacterial pathogens, improving the immune system, reducing the risk of colon cancer, lowering serum cholesterol levels, alleviation of lactose intolerance and nutritional enhancement [95]. Probiotics are commonly incorporated in dairy-based beverages that include many drinks; milk-based, yogurt, and cultured buttermilk. In 2008 to 2013, non-dairy products especially probiotic drinks emerged in the functional beverages market. Table 11.7 shows some of the examples of commercially available probiotic beverages [97, 147].

11.15 MINERALS

The human body needs minerals in order to build strong bones and muscles [149]. Besides, this inorganic compound is also important to heal the wound, repair cellular damage, and enhance the immune system [86]. There are two types of minerals which are major minerals (macrominerals) and trace minerals (microminerals). For major minerals, it comprises of sodium, potassium, chloride, calcium, phosphorus, magnesium, and sulfur. As for trace minerals, it consists of iron, zinc, iodine, copper, manganese, and fluoride [134]. Fruits and vegetables can be a good source of minerals thus can be the best functional ingredient in a beverage [133]. Beverages comprising of milk and fruit juices contain great sources of minerals. A study was done by Salvia-Trujillo et al., [149] has highlighted that by using High-Intensity Pulsed Electric Field (HIPEF) the mineral concentration in the fruit juices stayed highly stable during storage, regardless of the processing applied. The macromolecule calcium was the major mineral recognized in the fruit juice-whole and fruit juice-skimmed milk beverages. Hence, the HIPEF process is a possible treatment for advanced functional beverages with high elements of health correlated substances.

TABLE 11.7 Example of Commercially Available Probiotic Beverages

Brand	Producer	Probiotic content
Vita Bosa®	Biola Inc., Canada	*L. acidophilus, L. casei, B. bifidum*
Proviva®	Skane Dairy, Sweden	*L. plantarum* 299v
Gefilus®	Valio Ltd., Finland	*L. rhamnosus* GG
Goodbelly®	NextFood, USA	*L. plantarum* 299v
Whole Grain Probiotic Liquid®	Grainfields, Australia	*L. acidophilus, L. delbruekii, Saccharomyces cerevisiae var. boulardii*
Biola®	Tine BA, Norway	*L. rhamnosus* GG

In general, fermented coconut water consists of a significantly higher quantity of sodium as well as other minerals, such as magnesium, calcium, and potassium [103]. The high mineral content in coconut water has made it a popular substitute for sports drinks that contain high with electrolytes [83]. The probiotic fermentation does affect towards the mineral content formerly exists in the coconut water [103]. Their findings have shown that the nutritive advantages and hydrating properties of fermented coconut water can be estimated to be as good as those of the fresh coconut water. Normally, drinks that are specifically made for rehydration must be able to restore the volume of water lost and recover the electrolyte elements, particularly sodium. Sodium levels must be in between 20 to 30 mmol/l for a rehydration beverage to provide fluid replacement and to stimulate absorption [114]. Hence, this shows that sodium gives better hydration for the utmost people.

Maple syrup is one of the popular food and drink ingredients which is made of maple sap. Without pasteurization or sterilization process the maple sap is not suitable for human consumption. There are about seven minerals detected in the sterilized maple sap which can be applied as a functional beverage. This includes sodium (7.70±2.07), magnesium (1.14±0.64), aluminum (1.19±0.35), ferum (0.68±0.85), zinc (0.25±0.52), calcium (4.38±1.57), and potassium (5.74±2.90) calculated by mg/100g [179].

Functional coffee containing ginger has been reported to have an excellent source of selenium, zinc, manganese, copper, and iron but rather modest in sodium, calcium, potassium, and magnesium. With these properties have promoted a great competitiveness of the functional coffee compared to common coffee types that comprised of traditional, instant, and espresso [162]. In a pure origin coffee country that produces espresso beverages like Brazil, Ethiopia, Honduras, Guatemala, Papua New Guinea Colombia, India, Mexico, Kenya, Cuba, Timor, Mussulo, and China, a comparison of mineral characterization has been done. The results show that calcium and manganese were found to be the best chemical description of origin. Nonetheless, South American pure origin coffees are richer in the analyzed elements except for calcium, whereas samples from Central America have mostly lower mineral quantities, excluding for manganese [124]. According to Costa et al. [42], coffee silverskin which is the major coffee roasting by-product comprises large amounts of potassium (5 g/100 g), magnesium (2 g/100 g), and calcium (0.6 g/100 g). It has the ability to inhibit induced hemolysis and act in cell protection against oxidative injuries. This suggests that beverages made from coffee silverskin have a higher potential to be one of the functional beverages [42].

11.16 SUPERFRUIT EXTRACT

The term 'superfruit' is a comparatively recent term used in the food and beverage industries. It is considered a new marketing strategy in order to promote common or rare fruits which can be consumed as food or drinks, or used as ingredients in the production of functional foods, beverages, and nutraceuticals. Superfruits possessed high nutritional values due to their richness in nutrients, antioxidants, and taste appeal. It also has been proved to have potential health benefits to the consumers. Although, there are no scientific criteria for defining which fruits can be considered as superfruits and the scientific society did not come out with any specific description for "superfruits," however, antioxidants rich fruits that may possess various health benefits are likely to be included in this category [56]. Fruits such as pomegranate, blueberries, acaí, noni, acerola, and mangosteen are appropriate to be called as "superfruits" and can be obtained at fruit juices section in beverage market [56, 97].

Acai (*Euterpe oleracea*) is a type of palms that have an edible palm heart and the fruits are known as acai berry. Initially, acai berry has been consumed in the form of juice and pulp. It has a high antioxidant capacity and its extract has been incorporated as an ingredient in food and beverages. Several studies have demonstrated that acai berry possesses anti-cancer, hypocholesterolemic effects, and anti-inflammatory activity [56, 193]. Besides antioxidants, it also contains other nutrients such as vitamins, minerals, dietary fiber, and complex carbohydrates [185].

Another example of superfruit is pomegranate. Pomegranate (*Punica granatum* L.) is one of the oldest edible fruits and are usually consumed fresh as a whole fruit or transformed into juice, jellies, and coloring agents [74]. It contains important sources of antioxidants such as anthocyanin and the 3-glucosides and 3,5-diglucosides of delphinidin, cyanidin, and pelargonidin. Studies have shown that nutrients from the pomegranate possess antioxidant and antitumor activity, and can help to protect the body from ailments like heart problems, diabetes and osteoarthritis [63, 74, 185].

11.17 AMINO ACIDS

Amino acid is vital in practically all biological progressions in the human body [115]. This is because of amino acids give a structure to the cells and act as a building block of it [57]. A large number of cells, muscles, and tissue are made up of amino acids, meaning they perform many important bodily tasks such as carrying substances, combating pathogens, regulating processes, sending signals and act as a catalyst for biological reactions [173]. There are many types of amino acids for instance histidine, methionine, phenylalanine, threonine, isoleucine, leucine, lysine, tryptophan, and valine [19]. Human body can't produce its own amino acid and must be obtained through food or beverage that contains high with protein [158].

Amino acid level is a very important constituent to determine the quality of tea as a functional beverage. Mulberry leaves especially Morusalba are rich with amino acids specifically leucine, phenylalanine, lysine, histidine, arginine, threonine, valine, and methionine (Chan et al., 2016). In the Xinjiang province of China, Russian mulberry contains the highest essential amino acids followed by white mulberry and black mulberry [85]. That is why there are lots of mulberry tea products in the functional beverage market [68]. Another type of tea leaves that contain high amino acids is AnJiBaiCha albino mutant. The verdicts in this research have provided new insights that lead to the albino phenotype and amino acid enhancement of AnJiBaiCha white tea [178]. Kocadağli et al., [93] has found that brewing conditions intensely influence the presence of free amino acid in tea. There are 20 amino acids which comprise of 18 proteinogenic and 2 non-proteinogenic that were identified in green and black tea samples. After 2 min of brewing procedure, the quantity of free amino acids in tea infusion was 220.53 mg/l and 211.21 mg/l for black and green tea, respectively.

A normal milk beverage can be upgraded into a functional milk beverage by adding phenolic compounds (100 and 200 mg/l). The phenolic was extricated from olive vegetable water, and later on, being fermented with γ-aminobutyric acid (GABA)-producing (Lactobacillus plantarum C48) and human gastrointestinal (*Lactobacillus paracasei* 15N). During fermentation occur, the total concentration of free amino acids elevated from ca. 190±8.3 mg/l to 303.7±11.8–319.5±9.1 mg/l [152].

11.18 EMERGING TECHNOLOGIES IN FUNCTIONAL AND MEDICINAL BEVERAGES PROCESSING

11.18.1 HIGH-PRESSURE PROCESSING TECHNOLOGY

The preservation of fruit juices by thermal blanching above 80°C is a common process. Nevertheless, products subjected to these process conditions will typically lose their natural flavor as well as some of their nutrient content [144]. The development of a non-thermal approach such as High-pressure processing (HPP) is due to the market need which is a request for minimally processed foods [73].

High-pressure processing (HPP) or known as high hydrostatic pressure (HHP)is a non-thermal method that exposed foods to high pressure between 100 –1000 MPa for 1–20 minutes with the pressure-transmitting condition using water. Hydrostatic pressure is operated by heightening the free energy which can be accomplished through physical compression during pressure treatment in a closed system by decreasing the mechanical volume. HP processing is typically operated by a moderate increase in temperature (adiabatic heating), in which this temperature increase due to the composition of the food product [91, 92, 104, 117].

In general, two types of HPP devices are available, horizontal and vertical. The horizontal type of HHP tool is commonly used for commercial purposes since this type of design facilitates the loading and unloading of containers during the production process [108]. Figure 11.1 shows the horizontal type of HPP tool from Multivac Inc., German [76].

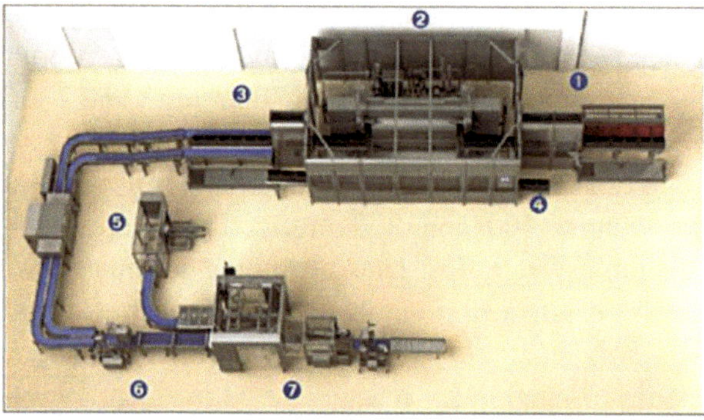

FIGURE 11.1 A horizontal HPP production systems.

HPP is able to preserve the nutritional quality of the product with minimal effect to quality which is associated with its little effect on the covalent bonds of low-molecular-mass compounds such as color and flavor compounds [25, 121]. HPP inactivates microorganisms and enzymes, therefore, it assists in enhancing the safety and shelf-life of perishable foods, whereas the thermal influence on nutritional and quality parameters are considered low in HHP since this procedure is operated at room temperature or moderately above [23, 25, 90].

In the treatment of HPP, individual phenolic compounds, total phenols, and the bioactivity of a food product regularly deteriorate. This might be happening because of the higher residual activity of polyphenol oxidase (PPO) and peroxidase (POD), two enzymes that are responsible for catalyzing the oxidation of the phenols. For example, these enzymes are proficient at oxidizing (+)-catechin, which causes deterioration, and then brown polymers will be generated [106].

Over the years, the observation of the application of HPP in the area of food and beverage processing has been done and many fruit juices products, such as HHP treated mandarin, grapefruit, apple, orange, carrot juices, and broccoli–apple juice mixture were found commercially in the market [26]. The advantages of HHP on the orange juice–milk beverage were observed in terms of nutrition, quality, and energy conservation [13]. The result showed that 5-log inactivation of *L. plantarum* in the orange juice–milk beverage had been achieved by the minimum HPP treatment. This minimum HHP treatment had shown a greater effect on sustaining the ascorbic acid level than the heat treatment. In another study by Barba et al. [14], blueberry juice subjected to HPP had shown that more than 92% of vitamin C retention and increased in total phenolic content. HHP also exerts the least influence on the quality of the treated juice. Nutrients and flavors of the product might be eliminated by the thermal processing, HPP benefited the quality of coconut water which the quality was observed to be similar to fresh coconut water [76]. In yogurt production, just before the fermentation, the HPP treatment had been applied to milk. As a result, HPP could produce fat set-type yogurt that contained 12% total solids with a creamy uniformity, without a need to add polysaccharides [167]. The additives used in the yogurt processing can influence the taste, flavor, aroma, and mouthfeel, however, HPP is advisable to be applied on the milk used for yogurt making as it could substitute the food additives [153].

11.18.2 MEMBRANE TECHNOLOGY

Membrane filtration technologies are increasingly employed in the functional beverage industry. The minimal beverage processing techniques could be generated by improving the product quality and stimulating energy-saving. Membrane processes technique benefited a low-cost and non-thermal separation effect as well as include no phase change or chemical agents, however, traditional juice processing technique demonstrated a distinctive performance. These benefits prompted an initiative of manufacturing new fruit juice containing natural fresh tastes and additives free initiates [28].

Membrane technology is a separation technique that involves a hydrostatic pressure gradient using semi-permeable membranes which depends on the specific permeability of the membrane for individual or mixed compounds of the fluid. In general, molecules in the ionic size range will be collected by reverse osmosis (RO) membranes, the dispersion of macromolecules are obtained by ultrafiltration (UF) membranes while micron size of particles is attained by microfiltration (MF). The definite osmotic pressure is applied to each of the membranes, RO, UF, and MF due to the membranes separated the molecules at a different size [29].

All three types of the membrane filtration involve a cross-flow filtration process (Figure 11.2). In this process, the liquid stream is treated at the feed section, and then it flows loosely throughout the membrane surface, therefore, two streams have occurred. Wolf et al. [175] described some particles in the solution will pass through the membrane and ready to be collected or the concentrated solids will be trapped at the membrane. Cross-flow filtration operates a self-cleaning which means the process needs extensive process times without cleaning compared to the traditional filtration and this is related to the feed stream that flows parallelly to the membrane not vertically [29].

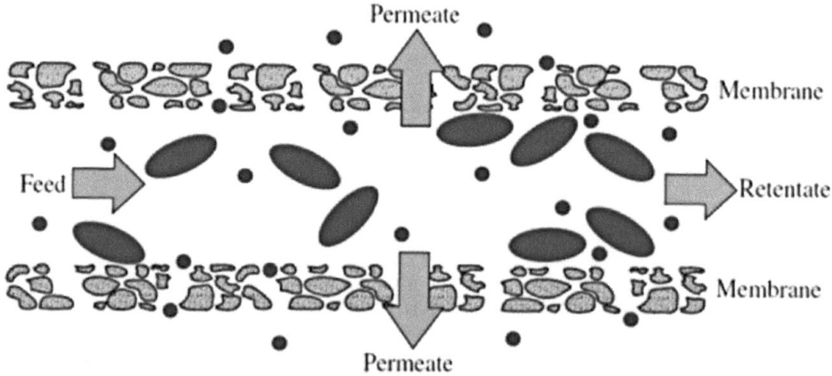

FIGURE 11.2 Cross-flow filtration process.

Ultrafiltration is commonly used in the juice industry to clarify and stabilize fruit juices and industrial-scale commercial processes have already been applied in the apple and orange juice industry [64]. Microorganisms, lipids, proteins, and colloids are a large species that could be collected through the UF technique, however, vitamins, salts, and sugars are a small compound that will flow together with water. Fukumoto et al. [59] had listed some of the advantages of fruit juice processed by UF rather than conventional technique such as increase juice yield, increase chances for single-step operating, reduce the chances of using gelatins, adsorbents and other filtration aids as well as an enzyme, easy cleaning and maintenance of the equipment, promote the exclusion of pasteurization, better juice clarity, reduce the filtration times and waste products. Besides, UF is also a commonly applied unit operation in dairy processing. It is used for the standardization of the protein content in milk.

Another membrane technology for juice clarification is microfiltration. Microfiltration (MF) is a membrane separation technique that involves a pressure system and the fruit juice can be clarified without the need for enzymatic treatment. A melon juice had been clarified using cross-flow MF and the juice was found to have a comparable quality with the fresh juice, but with an exception for insoluble solids and carotenoids [168]. Acerola juices were produced and clarified using MF, thus, the investigation towards consumer preferences had been done and 84% of consumers preferred the

MF juice [109]. Another investigation by Mirsaeedghazi et al. [113] had reported that the microfiltration is able to clarify pomegranate juice without encouraged significant chemical changes to its when compared to enzymatic methods. MF is also principally used in dairy processing to reduce the bacterial loading in a dairy fluid by retaining bacteria and spores. MF is a preferred method since it minimizes the use of heat treatment, which is beneficial for products which are heat sensitive to [101].

Reverse osmosis is a technique for concentrating liquid which is normally used in juice production. Fruit juices are concentrated to reduce water activity and also provide economic impact for storage, transportation, distribution, and commercial operations. RO operation allows the separation of water from the juice solids, however, it is restricted by high osmotic pressures. Therefore, the technique is applied for pre-concentration treatment that permits concentration values of about 30 g TSS (total suspended solids)/100 g, corresponding to osmotic pressures of about 50 bar [23]. No phase change involves in the RO process and this process is normally performed at room temperature which results in considerable energy saving to the industry. In addition, the RO process compact installations are also easily conducted. The RO quality of treated juice is better than the quality of the juice produced via conventional thermal treatment because heat treatment affects the sensory and nutritional properties of the juice (Álvarez et al., 2000). Furthermore, this method permits the retention of compounds by permeation of water and concentration of solutes such as salts and sugars [36] whereas the retention of water could be appropriate for recycling within the same plant [11, 61].

11.18.3 MICROWAVE HEATING TECHNOLOGY

Microwave heating technology is applied in many parts of the functional or medicinal beverage production such as in pre-treatment, extraction, and pasteurization or sterilization. In microwave heating, heat is being generated within the material, which is known as volumetric heating. This volumetric heating occurs due to dipolar rotation and ionic conduction of the constituents in the materials when exposing to the oscillating electric field in the microwave [170].

In conventional juice processing, the juice is commonly obtained via mechanical pressing. In order to maximize juice yield, in some fruit juice processes, enzymes are usually added into the fruit mash to assist cell walls degradation. This enzymatic cell wall degradation assists in increasing juice production. Alternatively, applying microwave heating on fruit mash before mechanical pressing also produces a higher yield of juice compared to mechanical pressing alone. Study conducted by Gerard and Roberts [62] on microwave heating of Fuji and McIntosh apple mash prior to the pressing process significantly increases apple juice yield. In this study, the apple mash was heated around 40°C to 70°C using a 2450 MHz microwave at 1500 W. The increase in juice production is due to the liberation of juice or bioactive compounds via cell wall breakage from rapid heating by microwave. Microwave heating destructive effect on orange peel tissue was demonstrated in the study by Kratchanova et al. [94].

Besides increasing juice yield, microwave technology has also been applied in the area of juice extraction particularly for 'hard to press' fruits. 'Hard to press' fruits are fruits that not suitable for mechanical pressing such as plum and apricot. This technology is known as microwave hydro-diffusion and gravity (MHG), wherein this technology, fruit juice is forced out via hydro-diffusion and the juice produced flow into the collector compartment via gravitational force (Cendres et al., 2011). No requirement of mechanical pressing and solvent and reduction in the number of juice processing steps are the advantages offered by MHG. MHG technique has successfully shown able to extract plum juice [31] and strawberry juice [166]. Even though the yield from microwave hydro-diffusion is low, additional pre-treatment methods such as enzymatic degradation help in increasing the juice yield.

Microwave-assisted extraction has long been used for the extraction of various plant bioactive compounds such as fruits, vegetables, spices, or agriculture by-products. Internal superheating caused by microwave energy absorption by water within the plant matrix promotes cell disruption subsequently improves the nutraceuticals recovery [172]. Microwave extraction of fruit juice by-product showed that more bioactive compounds can be extracted from fruit by-products, and then these compounds can be added into the juice to make the juice more valuable. Apple juice produced from microwave heating of apple mash has higher total phenolic and flavonoid content [62]. An innovative grape juice high in polyphenols was produced by Al Bittar et al. [5] using microwave-assisted extraction of grape by-products.

Another application of microwave technology in juice processing is in the area of pasteurization. Microwave has been studied as an alternative to the conventional pasteurization method. High temperature, pressure, and long processing duration in conventional pasteurization processes have been associated with a quality loss in fruit juice. Continuous flow microwave pasteurization has been proven to be effective in inactivating of E.coli and Listeria in apple juice [155]. Microwave pasteurization shows better retention of Kava juice quality compared to the conventional method [2]. Microwave pasteurization has shown to retain bromelain and vitamin c in pineapple juice [1]. In a comparative study between microwave and conventional pasteurization of tomato juice by Stratakos et al. [157], microwave pasteurized tomato juice showed to have similar physicochemical and microbiological properties as the conventional pasteurized juice, which means that microwave continuous system can be used to replace conventional pasteurization system since microwave treatment offers shorter processing time. Another plus point for microwave pasteurization is that the tomato juice produced via this technique has a higher antioxidant capacity, which is an added value to the product.

Studies in microwave heating also showed that microwave heating is capable of inactivating endogenous enzymes in fresh fruits and vegetables that cause degradation, polyphenol oxidase (PPO), and pectin methyesterase (PME). PPO is the enzyme responsible for the loss of quality in fresh fruits and vegetables while PME is the enzyme responsible for the loss of cloudiness in fresh juice [150]. PPO in PME activities in Kava juice was significantly reduced by microwave heating as given in Ref. [2].

11.18.4 SONICATION TECHNOLOGY

Sonication or ultrasound is a non-thermal process that can be applied in many parts of juice processing such as pre-treatment, extraction, and pasteurization. The sonication frequencies generally used in the power ultrasound region 20–25 kHz. The production and collapse of cavitation bubbles in the sonication process promote mechanical, chemical, and biochemical effects in the liquids [132]. The cavitation phenomenon occurs in sonication due to the interaction between ultrasonic waves, liquid medium, and dissolved gas in the liquid. Ultrasonic waves form bubbles and when the bubbles grow into unstable size, the bubbles will burst and release high heat and pressure that breaks the compounds in the liquid which induce particle dispersion and cell disruption effect into the system (Figure 11.3) [3]. The efficiency of the ultrasonic treatment can be increased by combining this treatment with temperature treatment, pressure treatment, and both temperature and pressure treatments, which are known as thermosonication, manosonication or manothermosonication, respectively [81].

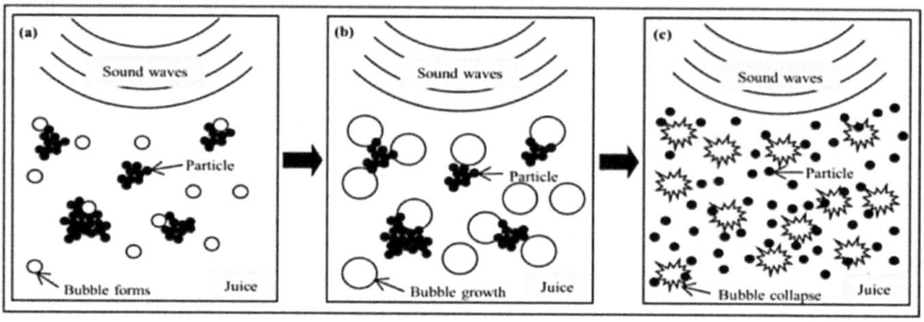

FIGURE 11.3 Cavitation phenomenon in liquid.

Beverage rich in valuable bioactive compounds can be obtained through ultrasonic extraction because the cavitation process in ultrasound treatment breaks down plant matrices to facilitate the extraction of bioactive compounds. The cavitation phenomenon during ultrasonic treatment assists in producing juice with a higher yield, better nutritional, and sensory quality with shorter processing time. Bora et al. [21] studied the effect of sonication on the banana juice yield. Higher juice yield was obtained from ultrasonic treatment combined with enzymatic treatment, compared to juice obtained from conventional extraction. In addition to that, the sonication technique in juice processing is demonstrated ability to increase the functional properties of the juice. This can be seen from the work by Abid et al. [4] where apple juice treated with ultrasound contained higher phenolic compounds and sugar contents when sonicated for 30 minutes and higher of the content of carotenoids and minerals after 60 minutes of sonication. The total phenolics, carotenoids, and retinol activity equivalent in ultrasonically treated Cape gooseberry juice were found to be higher than conventional extracted juice in the study conducted by Ordonez-Santos et al. (2017).

Besides increasing functional properties of the juice, with minimal loss of on sensory, color, and bioactive compounds of the juice. Significant loss of functional properties in the functional and medicinal drink is normally associated with the application of high temperatures during the conventional fruit drink production process. Sonication treatment offers treatment with a lower temperature than the conventional thermal process, therefore, it is able to retain the quality of the juice. A study performed by Khandpur and Gogate [87] on orange juice showed ultrasonically treated orange juice has a similar taste to the fresh untreated orange juice and it was accepted by consumers.

In addition, ultrasound treatment is also found to be effective in inactivating microorganisms at a lower temperature, therefore, ultrasonic is an alternative treatment for conventional pasteurization. Complete inactivation of microorganisms (total plate counts, yeast, and mold counts) in apple juice treated with thermosonication bath (25 kHz) and probe (20 kHz) at 60°C was achieved in a study done by Abid et al. [4]. Another study by Jambrak et al. [81] on thermosonication treatment of apple, cranberry and blueberry juices and nectars at temperature 60°C, amplitude between 60 to 120 um and exposure time between 3 to 9 minutes, demonstrated complete inactivation of Aspergillusochraceus 318, Penicilliumexpansum 565, Rhodotorula sp. 74 and Saccharomyces cerevisiae 5 in the juice samples.

Sonication is also used in juice processing to inactivate degrading enzymes polyphenol oxidase (PPO), polyphenol peroxidase (POD), and pectin methyesterase (PME). PPO and POD reactions cause discoloration and off-flavor in fruits and vegetable produce. Thermosonication using an ultrasonic bath (25 kHz) and probe (20 kHz) have been studied to inactivate PPO, POD, and PME in apple juice and the study result showed that these enzymes were completely inactivated using ultrasonic probe treatment at 60°C [4].

11.19 PHYSICAL AND CHEMICAL PROPERTIES OF FUNCTIONAL BEVERAGES

11.19.1 TEXTURE AND STABILITY

It is challenging to modify or create a beverage so that it can fulfill its claim of improving health or a functional property because the 'functional' ingredients added to achieve this aim may change the stability (e.g., proteins or minerals) or texture (e.g., fibers) of the beverage. For example, a beverage with reduced sugar levels to deliver on its promise of having 'no/lower/reduced' health claims are often affected by texturally and stability-wise. The very low viscosity matrix of the finished product further compound the challenges in product development. Generally, compared to more viscous products, beverages are not as robust when its formulation or processing is changed. It is crucial to consider the composition of the beverage when selecting the type of ingredients to add, as this directly impacts the stability and texture of the beverage. Besides that, other key factors affecting texture and stability include process parameters such as filling temperature, heat treatment (time, temperature), and homogenization. To offset the issues in the stability and texture of the final product, ingredients such as emulsifiers and hydrocolloids are often used.

11.19.1.1 FUNCTIONAL BEVERAGE TEXTURE AND STABILIZATION DEFINITION

Viscosity measurements are often used to measure or quantify the texture of a beverage. However, since viscosity is measured at a specific shear rate, this parameter only describes one attribute, but in reality, the texture is a very broad term—encompassing sensory descriptors and rheological properties as well. In other words, the texture is essentially the sensation and appearance of the beverage as a person drinks it, specifically:

- What does it look like when poured?
- What does the glaze look like on the bottle?
- What sensation does it give in the mouth at first sip?
- How does it coat the mouth?
- What does the residue look like?

According to Glicksman [65], the feel of the beverage in the mouth (i.e., mouthfeel) is largely dependent on the hydrocolloids (the flow properties of the gum), as it displays rheological behavior that affects the textural properties of gum. The term 'stabilization' is a broad, general term that covers various exceedingly complex bio-physicochemical mechanisms. Therefore, to address the issue of stability, each of its parameters must be categorized and described appropriately. In general, beverages that are smooth and homogenous are known as stabilized beverages. Stabilization can further be defined as follows, based on the targeted formulation and functionalities of the final beverage product:

- Stabilization of particles: the beverage must have an even suspension of particles (e.g., cacao particles, minerals, or pulps) throughout.
- Stabilization of Emulsion: the top of the beverage bottle or container must be devoid of fat or oil rings.
- Stabilization of Proteins: the beverage must exhibit a non-sandy and smooth feel in the mouth with no flocculation or sedimentation of proteins.

- Textural stabilization: the above functionalities all indicate texture stabilization. Besides, the beverage must have a homogenous-looking appearance and viscosity, specifically:
  - there are no lumps or gel points;
  - there is no gradient in viscosity;
  - there are no layers that have formed;
  - there are no separations in the phase;
  - there is no clarification;
  - there is no flocculation.

Depending on the process and formulation of the beverage, each parameter will require a different ingredient solution.

11.19.1.2 FUNCTIONAL BEVERAGE TEXTURE AND STABILIZATION MECHANISMS

11.19.1.2.1 Hydrocolloids as Gelling and Thickening Agents

Hydrocolloids mainly help thicken or gel beverages, so they are generally divided into gelling or thickening agents. A beverage with a thickening agent will have the desired texture but its particulates will not be suspended. According to Hoefler [75], thickening agents help to prevent particles from settling down too quickly and prevents oil droplets from rising rapidly, but separation will still occur. To address this issue, gelling agents are used, as they can prevent oil droplets or particles from separating by permanently suspending them in a matrix, by building a 3-D network through the formation of links between molecules. However, the yield value that the network creates must be higher than the density of the particulates for the agent to work effectively.

Acidic beverages contain food proteins such as casein that usually agglomerate and form sediments, so these proteins must be stabilized. However, applying heat to the proteins can dehydrate them easily and make them sandy. Therefore, specific hydrocolloids must be formulated to directly interact with the proteins via electrostatic interactions to prevent a chalky texture and sedimentation due to protein agglomeration. Complex coacervates can be formed when food proteins (isoelectric point around 5) react with anionic polysaccharides such as propylene glycol alginate (isoelectric point around 3.5) and high ester pectins, for example, carboxymethylcellulose (CMC) as long as the pH is below the isoelectric point of the protein and above the isoelectric point of the polysaccharide such that the reaction occurs in the intermediate region where the two macromolecules carry opposite net charges. Several factors determine the strength of the interaction between the polysaccharide and protein, namely the surface distribution of the ionizable groups, the three-dimensional structure of the protein, and the backbone distribution of the carboxyl groups. Futo [58] added that the presence of fat or sugars, pH, and ionic strength also affects the interaction as a whole. The flow curve of a beverage is an ideal approach for characterizing beverages.

Extensive research has shown that a shear rate ranging from 1–100 s^{-1} correlated to good stimuli when evaluating viscosity orally. Sworn [163] explained that the reason is that at this range, the flow behavior of the beverage and choice of hydrocolloid could be tweaked to meet specific sensory and processing requirements. However, Glicksmann [65] reported a much narrower shear range of 20–50 s^{-1}. Therefore, beverage development should always be evaluated based on its sensory attributes and flow curves. In this case, Glicksmann [65] highlighted two main benefits of conducting rheological measurements based on the beverage flow curves (i.e., viscosity versus shear rate): i) the best beverage out of several beverages containing different gums could be chosen based on the evaluation and characterization of the best mouthfeel using the flow curves, and ii) existing beverages can be qualitatively measured and characterized to be used as targets for duplication.

Nevertheless, because rheological methods require specific methods and rheometers, the process could be very time-consuming. Besides, because of their low viscosity and small particle content, beverages are extra difficult to rheologically evaluate. Therefore, besides rheological evaluation, sensory evaluation is also commonly carried out to characterize beverages. Because beverages tend to have low viscosities than other food matrices, their matrices have a polymer concentration that is around or a little above the critical concentration (C*). Therefore, the main challenge in formulating beverages is to ensure that there is minimal interaction between their polymers so that the particles in the 3-D network of the beverage are suspended, while still maintaining the low viscosity of the beverage.

11.19.1.2.2 Emulsifying Agents

The interfacial tension of two normally immiscible phases can be reduced with the use of emulsifiers. These substances prevent the creaming, coalescence, and flocculation of the different immiscible phases (that could lead to emulsion breakdown) during storage by allowing them to mix and form a stable emulsion. Becher and Walstra [16] observed that turbulent conditions almost always cause oil-in-water (o/w) emulsions to emulsify. Besides, compared to the contribution of the applied homogenization energy to oil droplet disruption, the effect of the interfacial tension is relatively negligible. Beverages that contain o/w emulsions such as juice beverages with essential oil-based or omega-3 oil-based flavors and high-fat milk beverages usually benefit from the use of emulsifiers, as these agents can stabilize the emulsion and ensure that it remains stable throughout its shelf-life.

11.20 SENSORY QUALITY

The sensory quality of functional beverages is assessed based on the properties and attributes listed below:

- Properties involving the use of organs (organoleptic properties): the consumer does not decide his beverage of choice based on tactile, auditory, visual, or olfactory elements, so these properties are not touched on in this section.
- Experiences after ingesting the drink (digestive properties): pleasure, fullness, and/or heaviness.

According to Hwang and Hong [78], the positive attributes that define a good quality beverage are tactile properties such as the absence of contaminants (strange flavors or odors), body, and mouthfeel; olfactory properties such as odor (orthonasal and retronasal) and aroma; taste properties such as aftertaste, mouth persistence, and flavor; and color and overall appearance. Meanwhile, negative attributes include gas, sedimentation, astringency and bitterness, discoloration, foaming, and unpleasant smell production (vinegary or ketone notes).

11.20.1 THE ACCEPTABILITY OF FUNCTIONAL BEVERAGES

The acceptability and availability of a beverage/drink amongst consumers are usually evaluated before it is released to the market. A hedonic scale is used to evaluate the attributes of a beverage, whether positive or negative. The scale usually varies from 5 to 11 points indicating maximum displeasure to maximum pleasure. For adults, verbal evaluation is used while facial hedonic scale evaluation is used for infants. The tests to measure whether or not a drink is accepted usually employ scales with reliability, good predictability, and discriminative power correlating to eating habits. Most tests employ the nine-point hedonic scale, but others also use seven-, five-, or three-point hedonic scales. Normally, acceptance tests involve consumers, where the level of confidence desired for decision-making will determine the number of participants involved. The commonly evaluated attributes of natural drinks and the associated sensory tests are listed in Table 11.8. Usually, the attributes of taste and smell and aroma and smell are evaluated based on the qualities mentioned below:

- The feelings that occur on the tongue and in the mouth such as sourness, bitterness, saltiness, or sweetness (and most recently umami) all fall under taste, indicating whether or not a drink is delicious.
- Drinks that evoke mixed and pleasant olfactory sensations or are perfumed or fragrant usually indicate a nice aroma.
- Usually, unpleasant olfactory sensations that refer to a specific type of aroma are indicated by the odor of the product.
- The viscosity or body of a drink indicates the oral sensation when ingesting the drink.
- Taste, viscosity, and aromatic sensations combine to denote flavor, such that the brain understands flavor by evaluating these components.

The time it takes for the flavor of a drink to disappear from the mouth is referred to as the flavor persistence in the mouth. This attribute is synonymous with wine tasting, but with the recent developments of juices and teas, this attribute has become an important one to evaluate. Usually, one's palate must be cleansed from bitter, fatty, spicy, or sweet tastes to allow for the experience of new sensory attributes. In this case, the ideal complement for cleaning the mouth is water. Since it is a given that water is used in the evaluation of sweet beverages and natural beverages, this combination of products is often not reported in publications.

11.20.1.1 FUNCTIONAL BEVERAGE STATISTICAL TESTS AND ANALYSES FOR SENSORY EVALUATION

Qualitative tests, discriminative tests, and acceptance tests are the tests used to evaluate the sensory attributes of beverages. The hedonic scale, namely Just About Right (JAR) and Labeled Affective Magnitude (LAM), are used in discriminative tests and acceptance tests. Meanwhile, the intensity grade scale is used in qualitative tests, to evaluate the taste profile and to conduct qualitative descriptive analysis (QDA) of the beverage.

TABLE 11.8 Attributes Commonly Evaluated in Natural drinks and the Sensory Tests Used

Product evaluated	Sensory evaluation	Sensory panel	Scale	Palate cleansing	Reference
Amarone red wine	Development of sensory descriptors for aroma, taste, flavor, and mouthfeel	12	Nine points	Undefined	[131]
Plum nectar (*Prunus domestica*)	Quantitative descriptive analysis (QDA): color, odor, taste, consistency and overall sensory impression	15	Intensity scale 0–10	Salt-free bread and water	[77]
Cloudy plum juices	Color, odor, taste, consistency and overall sensory impression	10–15	Nine points	Undefined	[102]
Black cherry, concord grape, and pomegranate juices blend	(a) Consumers: overall liking, appearance, Just about right (JAR) attributes (b) Descriptive analysis: development of sensory descriptors for flavor, mouthfeel and strange flavors	(a) 100 Consumers (b) 10	JAR scale nine points	Unsalted crackers and water	[99]
Arjuna-ginger medicinal mix blended	Appearance, color, taste, flavor, brightness, and strength	20	Nine points	Undefined	[171]
Apple and apple-whey beverages	(a) Appearance (color, sediments, and suspension); odor desirability, odor intensity, and flavor (b) Descriptive flavor analysis	(a) 15 (b) 18	Five points	Undefined	[82]
Grape juice mixed with apple, pear, and peach juices	Color, odor, aroma, sweet taste, persistence in the mouth, overall pleasantness	50 Consumers	Five points	Natural water	[38]
Mugwort tea (*Artemisia argyi* H. Lev & Vaniot)	Color acceptability, flavor acceptability, saltiness, bitterness, sourness, astringency, sweetness, and overall preference	15	Labeled affective magnitude (LAM) scale 15 points	Undefined	[80]

TABLE 11.8 *(Continued)*

Product evaluated	Sensory evaluation	Sensory panel	Scale	Palate cleansing	Reference
Drink lemon juice and honey ready to serve	Color/appearance, flavor/aroma, body, taste, and overall acceptability	Undefined	Nine points	Undefined	[154]
Custard apple leaf extract fortified sweetened aonla juice	Color and appearance, flavor, taste, and mouthfeel, overall acceptability	Undefined	Nine points	Undefined	[160]
Wine	Aroma, tactile attributes	8	Ten points	Water	[176]

The sequence of sensory tests used for beverage development is shown in the yellow box of Figure 11.4. Based on the optimization, substitution, or acceptance of ingredients in the test of acceptance, alternative tests are presented. Based on the review of the literature presented in this section, the sensory tests provide the researcher with different strength points to discriminate or accept a product. Hence, the questions that are to be answered during the beverage development product will determine whether or not the product is rejected or accepted.

Consumer acceptance tests normally use a nine-point verbal hedonic scale indicating degrees of taste from "extremely disliked" to "extremely liked." Following that, the verbal responses are transformed into numerical values to further analyze statistically. A study attempted to discover whether the words on the nine-point hedonic scale could be changed to numbers [119]. They found that scales that used "only words" or "only numbers" usually induced the customers to respond differently. In particular, 79% to 100% of the customers would give different results when given either scale. The study concluded that the researcher should apply caution when transforming one scale to the other, as the numerical data derived from these two scales are not interchangeable. Lawless et al. [100] mentioned the LAM scale, which is another scale used in consumer acceptance tests that have higher discriminative power than the nine-point scale. The LAM scale also has good spacing between anchor words for foods and well-liked beverages. Lawless et al. [100] further compared the nine-point hedonic scale and the LAM scale in food acceptance tests. They found that both discriminated products well based on the tastes of the consumer. Also, they found that consumption patterns had a strong effect on consumer acceptance ratings. The study concluded that the nine-point hedonic scale was comparable to the LAM scale. They suggested using more comprehensive scales to ensure that the true differences between products could be detected, and thus preventing Type II error.

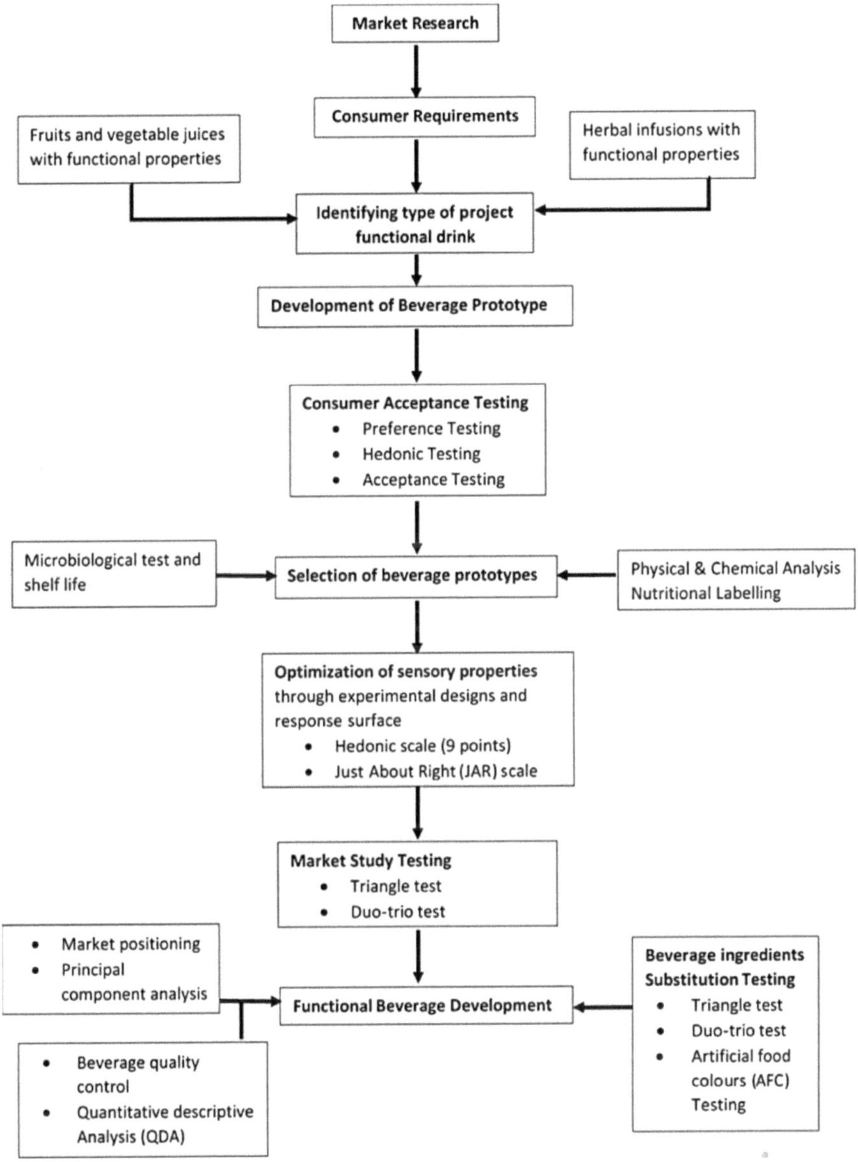

FIGURE 11.4 Sequence of sensory tests that are used in the development of natural beverages.

The LAM scale was used in a previous study to determine the acceptability of two species of Artemis tea based on general preferences, bitterness, acidity, color, taste, and salinity [80]. The range for the LAM scale was 0 to 15, correlating to the greatest imaginable dislike and greatest imaginable like. The study found that the differences in the volatile compounds in the tea, namely the terpenic compounds, contributed to significant differences in general preferences and taste acceptability while the rest of the attributes showed no significant difference.

The JAR scale is normally used to identify attributes that need improvement or to optimize a product attribute. The scale shows that the ideal value of an attribute is close or near to the most-liked attribute. In other words, after evaluation using the JAR scale, products that qualify as 'just right' products indicate the most-preferred or most-liked products. Nevertheless, studies that have employed the JAR scale often discover very different optimal values of attributes compared to the attributes of current products on the market.

For example, to optimize the sweetness of lemonade, Epler et al. [52] assessed the box and line JAR scales against the hedonic scale. Several formulations of the lemonade were developed with different sugar content (6% to 14%) and the taste and the "optimal" level of sweetness of each were predicted. Both the line JAR scale and the box JAR scale produced similar optimum values (9.2% and 9.4% sucrose), but this was lower than the results of the hedonic scale, which showed 10.3% sucrose as the optimum sweetness. Then, a preference test was conducted, showing that consumers preferred the lemonade with 10.3% sugar content compared to the 9.3% lemonade. Therefore, the JAR scale is not as good at predicting sweetness as the hedonic scale.

The optimal sweetness concentration of a beverage could be assessed using 3- and 9-point JAR scales. The data from both scales were compared via a survival statistical analysis and then a regression analysis. The study used orange juice with three different sucrose concentrations. It was found that the 3-point JAR scale showed an optimum sucrose concentration of 13.1%, higher than the 8.2% optimum sucrose concentration produced by the nine-point scale. Later, additional tests showed that the sample with 13.1% sucrose concentration was preferred by 70% of subjects rather than the sample with 8.2% concentration. The study, therefore, showed that combining statistical analysis with the 3-point JAR scale yielded more real optimal sucrose concentration compared to the 9-point scale. The Box and Wilson methodology (also known as response surface) have also been used to optimize drinks. This method combines regression analysis with experimental design. This method enables the determination of the optimal value of an independent variable (e.g., sugar level) that could maximize the response variable (e.g., flavor). One study used the response surface methodology to find the optimal level of sweetness for a soybean extract fermented beverage. They obtained predictive response models that indicated the ideal sweetness, flavor, and generally accepted purchase attitude, with the ideal sweetness being 11 g of sucrose per 100 ml.

Elsewhere, Pagliarini et al. [131] differentiated wines from different harvests based on their taste and aroma, also with the help of response surface methodology. A few studies have conducted similar tests, but the above studies better illustrated the use of the method for developing beverages. Table 11.9 compares the various different tests that have been developed [54].

TABLE 11.9 Comparison Between Different Sensorial Tests

Test	Samples	Presentation sequence	Judge instruction
Triangular	Encoded with three-digit numbers. Two samples are the same and are presented with different codes. The third sample is different.	AAB ABA BAA BBA BAB ABB	Which sample is different from the other two?
2AFC	Encoded as A or B	AB and BA	What is the sweetest sample? (for example)
3AFC	Two samples are the same and are marked with three-digit codes. The third sample is the one that has the greatest strength in the attribute and is coded in the same way	ABB, BAB, BBA, BAA, ABA, and AAB	What is the saltiest sample? What is the least salty sample?
Duo-trio	A sample is marked as R = reference and the other samples are marked as A or Bs	(a) Constant reference: R AB, R BA (b) Balanced reference: RA AB, RA BA, RB AB, RB BA	(a) Select the sample similar to A (b) Select the sample similar to A or B

Difference tests are usually used to compare the attributes of a new beverage or to replace the ingredient in a beverage. The differences tests are divided into two categories: (i) duo-trio and triangular tests (when the cause of the difference is not asked, and other attributes are focused on), and (ii) 2AFC and

3AFC tests (when the cause of the difference is asked, focusing on a specific attribute). Normally, these tests are conducted on consumers and trained judges. Simpler tests are conducted with consumers where the reliability of the results can be increased with an increase in the number of participants.

The Chi-square statistical method is used in the differences test. Because this test requires repeated sample testing such that adaptation to stimulus could occur, the judge can become fatigued. Hence, one session only allows the evaluation of six samples at most. The results of this test are interpreted using statistical tables, depending on the required significance level, sample size, and the minimum number of correct answers. A new drink that is about to launch in the market must first have an established flavor profile, defined based on its sensory properties. Besides, the properties must also be monitored in the process of quality control. In this case, the sensory attributes of the product, such as appearance, aftertaste, and mouthfeel, but no more than seven attributes, all in all, can be determined via a quantitative descriptive analysis (QDA) involving 10–12 trained judges. The QDA is conducted to quantitatively specify and establish the nature and intensity of a product's sensory attributes.

In a QDA, the intensity of the attributes of some beverages is evaluated based on the intensity level on a vertical line. Once converted into numerical values, the distances between the vertical lines are analyzed using Analysis of Variance (ANOVA). In one study, a QDA was conducted on nectars and 10 descriptive terms were generated to describe the products, relating to consistency, smell, taste, color, and overall sensory impression [77]. Based on the ANOVA results, the overall sensory impression, sour and sweet intensity, and color intensity of the nectars were identified as significantly different, but the addition of acid or sugar did not yield any significant differences.

To establish the position of a beverage in the market, Principal Component Analysis (PCA) can be used. This analysis leverages on the beverage's sensory attributes, namely body, color, flavor, and aroma. Using PCA, the sensory attributes of a beverage could be quantitatively analyzed based on the correlation between the attributes. It also enables the calculation of new variables by grouping attributes such that the distance between groups and attributes can be observed. Therefore, the product with better market positioning among consumers can be identified, and its consequences on the market assessed.

11.21 FLAVOR RETENTION AND RELEASE

Flavor retention and release are mostly studied in functional beverages (including drinkable meal replacers or sports supplements), to design healthier food products (low-fat milk, alcohol-free beer, etc.) without compromising on traditional product acceptability, and beverages with exotic features (exotic fruit tastes, cocktails, fusions, etc.). In skim milk, loss/lack of hydrophobic flavors challenges consumer's acceptability compared to that of high-fat milk, whereas the potential health benefit of soy milk suffers from a beany off-flavor [89] originating from lipoxygenase activity [107]. Flavor release or retention is generally affected by the intrinsic chemical properties of the flavor (hydrophobicity, hydrophilicity (log P value), and volatility), the composition of the medium (lipid, protein, salt, sugar, etc.), and finally environmental conditions (temperature, pH). In other words, the interaction between flavor compounds and other food ingredients under given environmental conditions determines the intensity of flavor retention or release from a product.

11.22 QUALITY AND SAFETY ISSUES IN FUNCTIONAL AND MEDICINAL BEVERAGES

Maintaining bioactivity of the active ingredient and quality of the functional and medicinal beverages after the processing stage is very important. It has always been a challenge to provide a fresh and perfect functional beverage to the consumers because as the time goes by, the quality of the product may deteriorate and their functionality may reduce or lost due to several reasons, such as degradation of the functional ingredient during production, short shelf-life and unsuitable packaging and storage condition [7].

11.22.1 SHELF LIFE OF FUNCTIONAL AND MEDICINAL BEVERAGES

During the extended storage of the functional and medicinal beverage in refrigerated conditions, the changes in quality of the product especially in the sensory and nutritional value are unavoidable. These changes limit the shelf-life of the product. According to Rysstad and Johnstone [146], the changes in the quality of certain beverages can be due to; (i) the activity of enzymes or microbes that are present in the packaged product, (ii) chemical changes in the product such as oxidation, and (iii) interactions of the beverage with the packaging material. A huge challenge is faced by the researcher in improving the shelf life of functional beverages. Therefore, many investigations are being done to enhance the viability and the efficacy of functional ingredients used in the beverage as it should be sustained in the food product until it reaches the end-user. One of the methods includes the usage of electromagnetic heating such as ohmic heating due to electrical resistance (50–60 Hz); radio frequency (RF) heating (10–60 MHz) and microwave heating (1–3 GHz) [146, 180]. Conventionally, heat or thermal treatment such as low-temperature long time (LTLT) and high-temperature short time (HTST) has been used in pasteurization to eliminate microbial contamination in the beverage [145]. Although this treatment able to be used to extend the shelf life by reducing the activity of the enzymes present in the product, however, it can cause adverse reactions such as nutrient degradation and reduce the freshness of the beverage [10, 145]. Therefore, in order to overcome this problem, electromagnetic heating has been employed by transferring the direct heat from the source into the food without going through a heat transfer surface. These methods are implemented homogeneous and quick of heating and cooling to promote less damage from thermal treatment thus increase the efficacy of the functional beverages.

According to the research done by Benlloch-Tinoco et al., [17] microwave pasteurization of formulated fruit juice is expected to bear a longer shelf-life than formulated fruit juice treated by conventional thermal pasteurization. This is due to greater penetration depths of 2450 MHz microwave radiation, which results in the better inactivation of spoilage microorganisms present in fruit juices. An excellent quality of fruit-based products might be accomplished through microwave heating due to greater potentials in prolonging the shelf-life which then can substitute many conventional preservation methods. This method provides faster heating rates, higher thermal efficiency, higher penetrative power, and shorter processing times due to the specific way of heating that takes place during microwave processing, thus results in better organoleptic, functional, and nutritional properties preservation, especially on the color.

Besides that, natural preservatives such as bacteriocins, organic acids, lactoperoxidase, essential oils, carotenoids, and phenolic compounds have shown a great potential in extending the shelf-life [8, 145]. For instance, natural beverage preservatives could be produced from fruit by-products containing phenols and carotenoids as they are able to prolong the shelf-life of the product by slowing down the occurrence of off-flavors and rancidity. Besides, they also exhibit well-known health benefits (Carbo et al., 2014). Other than electromagnetic heating and natural preservatives, there is also another method that can be used to preserve and prolong the shelf life of the beverage such as pulsed electric fields [130], reverse osmosis and ultrasonication [17].

11.22.2 STABILITY OF FUNCTIONAL AND MEDICINAL BEVERAGES

The suitable concentration of a functional ingredient incorporated into a beverage to achieve the optimum efficacy of the specific functional compound are often very challenging as these functional compound may directly influence on the physical properties and the stability of the product. Besides that, process parameters such as homogenization, filling temperature, and heat treatment such as time and temperature also play an important role to provide stability to the final product.

Commonly, a stabilized beverage defines a well uniform and smooth beverage. Referring to the beverage formulation and specific final functionalities, stabilization can be particularly defined into three categories; (i) particle stabilization where the particles such as pulps, cacao particles, and minerals are uniformly suspended in the beverage, (ii) emulsion stabilization in which fat or oil ring does not appear on the top of the beverage packaging, and (iii) protein stabilization where no flocculation or sedimentation of proteins occurs in the beverage or the beverage exhibits a smooth and non-sandy mouthfeel [54]. To overcome these challenges, the use of ingredients such as hydrocolloids, polyphenols, emulsifiers, and enzymes can help in ensuring the stability and texture of the final product. [7, 54].

11.23 STORAGE AND PACKAGING

The quality and the longevity of the final product can be directly or indirectly affected by the packaging material and the storage condition [7]. The choice of the packaging materials can influence the interaction between the product and its environment. There are four types of interaction that are involved which are (i) egress and scalping that causes aroma loss, (ii) sorption, (iii) migration of monomers and additives from the packaging materials; and (iv) ingress of oxygen and light (Figure 11.5).

Functional beverages can also be damaged when exposed to light. For example, bioactive compounds such as antioxidants, vitamins, or other light-sensitive compounds that incorporated in functional beverages must be shielded from light exposure functional drinks may contain special antioxidants, vitamins, or other light-sensitive ingredients that must be protected from light exposure during storage. When the products are exposed to the light, it may result in accelerated vitamin C degradation as well as photobleaching [146]. To overcome this, aluminum-foil lined cartons can be used to have a complete light protection.

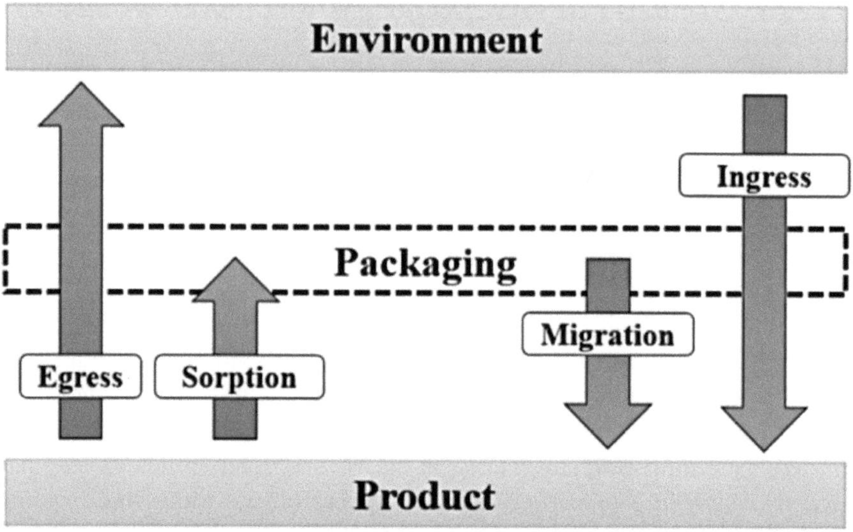

FIGURE 11.5 Influence of food-packaging interaction on the product shelf-life.

Another method that can be used to improve the quality of the product is by sterilization of the packaging. The most common packaging utilizes the combination of concentrated hydrogen peroxide, heat, and UV treatment in cartons which encourage bacterial spores to be inactivated synergistically and then able to produce a sterilized condition [9, 146]. Another potential method in the sterilization of product packaging includes pulsed light [66], plasma, and UV-excimer [174].

Another important factor that needs to be considered in ensuring the best condition of the finished product is the storage temperature. A study was done by Andersson et al., [7] shows that storage temperature influenced the quality of bioactive compounds in the beverage. Higher storage temperature (20°C) causes rapid changes in the level of the bioactive component as compared to lower storage temperature (4°C). Therefore, it is crucial for a rapid cooling and lower storage temperature to be applied to the product after the packaging had been done in order to preserve the functional nutrients.

11.24 CONCLUSIONS

Food and beverage manufacturers are continuously facing challenges in producing new formulations and development of new beverages especially beverages with added functional ingredients that could promote health. The challenges include the quality and safety aspects of the beverages while maintaining or improving the taste for consumer's benefit. Functional beverages that could benefit health, improve energy and immunity system especially that are relatively easy to consume would be very high in demand as the consumers are looking more for convenient due to busy lifestyle. Furthermore, before promoting its health-promoting effect to the consumers, a quality functional ingredient should be analyzed and tested for clinical evidence for its safety and efficacy. It is easy to claim that the products are having a health-promoting effect but it would be unfair for consumers to get the side effect if the issues of safety are not taken seriously by the manufacturers. The market trend in functional and medicinal beverages is increasing and it would be beneficial for the manufacturers and researchers to collaborate in improving the quality and producing more innovative products in the future.

ACKNOWLEDGMENTS

The authors would like to acknowledge the Ministry of Education Malaysia and Universiti Teknologi Malaysia for the financial support (Research Grant No. R.J130000.7851.4F993 and Q.J130000.3551.05G90). Special thanks to Nurul Hazirah Hamidon, Nabilah Abdul Samad, Farhani Ahmad, Chin Chooi Li, and Hazelynna Makerly for the contributions in this chapter.

KEYWORDS

- **functional beverages**
- **healthy drinks**
- **processing**
- **quality**
- **safety issues**

REFERENCES

1. Abd Aziz, N. A., Jusoh Mohd Jusoh, Y. M., Nik Mahmood, N. A., Yunus, N. A., & Endut, A. (2017). Quality evaluation of microwave and conventional pasteurized

pineapple juice. *International Food Research Conference Proceeding, Universiti Putra Malaysia.*

2. Abdullah, S. A., Lee, S. H., Cho, I. K., Li, Q. L., Jun, S., & Choi, W. (2013). Pasteurization of Kava juice using novel continuous-flow microwave heating technique. *Food Science and Biotechnology, 22*(4), 961–966.
3. Abdullah, N., & Chin, N. L. (2014). Application of thermosonication treatment in processing and production of high quality and safe-to-drink fruit juices. *Agriculture and Agricultural Science Procedia, 2*, 320–327.
4. Abid, M., Jabbar, S., Wu, T., Hashim, M. M., Hu, B., Lei, S., & Zeng, X. (2014). Sonication enhances polyphenolic compounds, sugars, carotenoids and mineral elements of apple juice. *Ultrasonic Sonochemistry, 21*, 93–97.
5. Al Bittar, S., Perino-Issartier, S., Dangles, O., & Chemat, F. (2013). An innovative grape juice enriched in polyphenols by microwave-assisted extraction. *Food Chemistry, 141*, 3268–3272.
6. Andersson, S. C., Ekholm, A., Johansson, E., Olsson, M. E., Sjoholm, I., Nyberg, L., Nilsson, A., & Rumpunen, K. (2015). Effect of storage time and temperature on the stability of bioactive compounds in aseptically packed beverages prepared from rose hips and sea buckthorn berries. *Agriculture and Food Science, 24*, 273–288.
7. Aneja, K. A., Dhiman, R. D., Aggarwal, N. K., & Aneja, A. (2014). Emerging preservation techniques for controlling spoilage and pathogenic microorganisms in fruit juices. *International Journal of Microbiology*, 1–14
8. Ansari, I. A., & Datta, A. K. (2003). An overview of sterilization methods for packaging materials used in aseptic packaging systems. *Food and Bioproducts Processing, 81*(1), 57–65. https://doi.org/10.1205/096030803765208670
9. Awuah, G. B., Ramaswamy, H. S., & Economides, A. (2007). Thermal processing and quality: Principles and overview. *Chemical Engineering and Processing: Process Intensification, 46*, 584–602.
10. Balannec, B., Vourch, M., Rabiller-Baudry, M., & Chaufer, B. (2005). Comparative study of different nanofiltration and reverse osmosis membranes for dairy effluent treatment by dead-end filtration. *Separation and Purification Technology, 42*(2), 195–200.
11. Bae, Y. W., Lee, S. M., & Kim, K. O. (2016). Age and gender differences in the influence of extrinsic product information on acceptability for RTD green tea beverages. *Journal of Science and Food Agriculture, 96*(4), 1362–1372.
12. Barba, F. J., Jäger, H., Meneses, N., Esteve, M. J., Frígola, A., & Knorr, D. (2012). Evaluation of quality changes of blueberry juice during refrigerated storage after high-pressure and pulsed electric fields processing. *Innovative Food Science and Emerging Technologies, 14*, 18–24.
13. Barba, F. J., Esteve, M. J., & Frigola, A. (2013).Physicochemical and nutritional characteristics of blueberry juice after high-pressure processing. *Food Research International, 50*(2), 545–549.
14. Bech-Larsen, T., & Grunert, K. G. (2003). The perceived healthiness of functional foods: A conjoint study of Danish, Finnish, and American consumers' perception of functional foods. *Appetite, 40*(1), 9–14.
15. Becher, P., & Walstra, P. (1983). In *Encyclopedia of Emulsion Technology*, Marcel Dekker Inc., 57.

16. Benlloch-Tinoco, M., Igual, M., Rodrigo, D., & Martínez-Navarrete, N. (2015). Superiority of microwaves over conventional heating to preserve shelf-life and quality of kiwifruit puree. *Food Control*, *50*, 620–629.
17. Bigliardi, B., & Galati, F. (2013). Innovation trends in the food industry: the case of functional foods. *Trends Food Science and Technology*, *31*, 118–129.
18. Bihuniak, J. D., & Insogna, K. L. (2015). The effects of dietary protein and amino acids on skeletal metabolism. *Molecular and Cellular Endocrinology*, *410*, 78–86.
19. Bomkessel, S., Broring, S., Omta, S. W. F., & van Trijp, H. (2014). What determines ingredient awareness of consumers? A study on ten functional food ingredients. *Food Quality and Preference*, *32*, 330–339.
20. Bora, S. J., Handique, J., & Sit, N. (2017). Effect of ultrasound and enzymatic pre-treatment on yield and properties of banana juice. *Ultrasonics Sonochemistry*, *37*, 445–451.
21. Boroski, M., Giroux, H. J., Sabik, H., Petit, H. V., Visentainer, J. V., Matumoto-Pintro, P. T., & Britten, M. (2012). Use of oregano extract and oregano essential oil as antioxidants in functional dairy beverage formulations. *Journal of Food Science and Technology*, *47*, 167–174.
22. Boye, J. I. (2015). Nutraceutical, and functional food processing technology. *Statewide Agricultural Land Use Baseline*, *1*.
23. Brown, P. N., & Chan, M. (2009). An overview of functional food regulation in North America, European Union, Japan, and Australia. In J. Smith., & E. Charter (Eds.), *Functional Food Product Development* (pp. 257–292). New York, NY: John Wiley & Sons.
24. Butz, P., Fernández García, A., Lindauer, R., Dieterich, S., Bognár, A., & Tauscher, B. (2003). Influence of ultra high pressure processing on fruit and vegetable products. *Journal of Food Engineering*, *56*, 233−236.
25. Buzrul, S., Alpas, H., Largeteau, A., & Demazeau, G. (2008). Modeling high-pressure inactivation of *Escherichia coli* and *Listeria innocua* in whole milk. *European Food Research International*, *227*, 443−448.
26. Campbell, B. (2013). Dietary carbohydrate strategies for performance enhancement. In B. Campbell (Ed.), *Sports Nutrition–Enhancing Athletic Performance* (pp. 75–124). Florence, SC: Taylor & Fran.
27. Cassano, A., Conidi, C., & Drioli, E. (2011). Clarification and concentration of pomegranate juice (Punica granatum L.) using membrane processes. *Journal of Food Engineering*, *107*(3–4), 366–373.
28. Cassano, A., Conidi, C., & Drioli, E. (2014). Membrane processing. In *Conventional and Advanced Food Processing Technologies*, 537–566.
29. Cappelletti, S., Daria, P., Sani, G., & Aromatario, M. (2015). Caffeine: Cognitive and physical performance enhancer or psychoactive drug?. *Current Neuropharmacology*, *13*(1), 71–88.
30. Cendres, A., Chemat, F., Page, D., Le Bourvellec, C., Markowski, J., Zbrzezniak, M., Renard, C. M. G. C., & Plocharski, W. (2012). Comparison between microwave hydrodiffusion and pressing for plum juice extraction. *LWT-Food Science and Technology*, *49*, 229–237.
31. Chacko, S. M., Thambi, P. T., Kuttan, R., & Nishigaki, I. (2010). Beneficial effects of green tea: A literature review. *Chinese Medicine*, *5*(13), 1–9

32. Chauhan, B., Kumar, G., Kalam, N., & Ansari, S. H. (2013). Current concepts and prospects of herbal nutraceutical: A review. *Journal of Advanced Pharmaceutical Technology and Research*, *4*(1), 4–8.
33. Chen, K. I., Erh, M. H., Su, N. W., & Chou, C. C. (2012). Soyfoods and soybean products: from traditional use to modern applications. *Applied Microbiology and Biotechnology*, *96*, 9–22.
34. Chen, Z. Y., Huang, Y., Leung, F. P., Tian, X. Y., Wong, W. T., Yung, L. M., Yung, L. H., & Yao, X. Q. (2008). Tea polyphenols benefit vascular function. *Inflammopharmacology*, *16*, 230–234.
35. Cheryan, M. (1998). *Ultrafiltration, and Microfiltration Handbook*. Lancaster, USA: Technomic Publishing Co Inc.
36. Childs, J. L., Yates, M. D., & Drake, M. A. (2007). Sensory properties of meal replacement bars and beverages made from whey and soy proteins. *Journal of Food Science*, *72*, S425–S434.
37. Chiusano, L., Cravero, M. C., Borsa, D., Tsolakis, C., Zeppa, G., & Gerbi, V. (2015). Effect of the addition of fruit juices on grape must for natural beverage production. *Italian Journal of Food Science*, *27*, 375–384.
38. Committee on Nutrition and the Council on Sports Medicine and Fitness. (2011). Clinical report—sport drinks and energy drinks for children and adolescents: Are they appropriate?. *Pediatrics*, *127*, 1182–1189.
39. Corbo, M. R., Bevilacqua, A., Petruzzi, L., Casanova, F. P., & Sinigaglia, M. (2014). Functional beverages: The emerging side of functional foods. *Comprehensive Reviews in Food Science and Food Safety*, *13*(6), 1192–1206.
40. Costa, A. S. G., Alves, R. C., Vinha, A. F., Costa, E., Costa, C. S. G., Nunes, M. A., Almeida, A. A., Santos-Silva, A., & Oliveira, M. B. P. P. (2018). Nutritional, chemical and antioxidant/pro-oxidant profiles of silverskin, a coffee roasting by-product. *Food Chemistry*, *267*, 28–35.
41. Das, L., Bhaumik, E., Raychaudhuri, U., & Chakraborty, R. (2012). Role of nutraceuticals in human health. *Journal of Food Science and Technology*, *49*(2), 173–183.
42. Deng, Y., Misselwitz, B., Dai, N., & Fox, M. (2015). Lactose intolerance in adults: Biological mechanism and dietary management. *Nutrients*, *7*(9), 8020–8035.
43. Dhingra, D., Michael, M., Rajput, H., & Patil, R. T. (2012). Dietary fiber in foods: A review. *Journal of Food Science and Technology*, *49*(3), 255–266.
44. Diplock, A. T., Aggett, P. J., Ashwell, M., Bornet, F., Fern, E. B., & Roberfroid, M. B. (1999). Scientific concepts of functional foods in Europe: Concensus document. *British Journal of Nutrition*, *81*(Suppl. 1), S1–S27.
45. Dolan, L. C. (2011). Claims: The United States and European perspective. *Agro Food Industry Hi-Tech*, *22*, 5–7.
46. Duncan, M. J., & Hankey, J. (2013). The effect of a caffeinated energy drink on various psychological measures during submaximal cycling. *Physiology and Behavior*, *116–117*, 60–65.
47. Elvin-Lewis, M. (2001). Should we be concerned about herbal remedies? *Journal of Ethnopharmacology*, *75*(2), 141–164.
48. Ernst, E. (1998). Harmless herbs? A review of the recent literature. *The American Journal of Medicine*, *104*(2), 170–178.
49. Epler, S., Iv, E. C., & Kemp, K. E. (1998). Hedonic scales are a better predictor than scales of optimal sweetness in lemonade. *Journal of Sensory Studies*, *13*, 191–197.

50. Fallourd, M. J., & Viscione, L. (2009).Ingredient selection for stabilization and texture optimization of functional beverages and the inclusion of dietary fiber. In P. Paquin (Ed.), *Functional and Speciality Beverage Technology*, 3–38.
51. Farbstein, D., Kozak-Blickstein, A., & Levy, A. P. (2010). Antioxidant vitamins and their use in preventing cardiovascular disease. *Molecules*, *15*(11), 8098–8110.
52. Felzenszwalb, I., da Costa Marques, M. R., Mazzei, J. L., & Aiub, C. A. (2013). Toxicological evaluation of Euterpeedulis: A potential superfruit to be considered. *Food and Chemical Toxicology*, *58*, 536–544.
53. Foegeding, E. A., Plundrich, N., Schneider, M., Campbell, C., & Lila, M. A. (2017). Protein-polyphenol particles for delivering structural and health functionality. *Food Hydrocolloids*, *72*, 163–173.
54. Futo, l. (1993).Pectin food use. In *Encyclopedia of Food Science and Nutrition* (2nd ed.) (Vol. 7, pp. 4445, 4448, 4449). Benjamin Cabillero.
55. Fukumoto, L. R., Delaquis, P., & Girard, B. (1998). Microfiltration, and ultrafiltration ceramic membranes for apple juice clarification. *Journal of Food Science*, *63*, 845–850.
56. Gaanappriya, M., Guhankuma, R. P., Kiruththica, V., Santhiya, N., & Anita, S. (2013). Probiotication of fruit juices by Lactobacillus acidophilus. *International Journal of Advanced Biotechnology and Research*, *4*, 72–77.
57. Galambos, I., Mora, J., Járay, P., Vatai, G., & Bekássy-Molnár, E. (2004). High organic content industrial wastewater treatment by membrane filtration. *Desalination*, *162*, 117–120.
58. Gerard, K. A., & Roberts, J. S. (2004). Microwave heating of apple mash to improve juice yield and quality. *LWT-Food Science and Technology*, *37*(5), 551–557.
59. Gil, M. I., Tomas-Barberan, F. A., Hess-Pierce, B., Holcroft, D. M., & Kader, A. A. (2000). Antioxidant activity of pomegranate juice and its relationship with phenolic composition and processing. *Journal of Agricultural and Food Chemistry*, *48*, 4581–4589.
60. Girard, B., & Fukumoto, L. R. (2000). Membrane processing of fruit juices and beverages: A review. *Critical Reviews in Food Science and Nutrition*, *40*, 91–157.
61. Glicksman, M. (1982). Functional properties of hydrocolloids. In *Food Hydrocolloids* (Vol. 1, pp. 59, 61). Boca Raton, FL: CRC Press Inc.
62. Gomez-Lopez, V. M., Ragaert, P., Debevere, J., & Devlieghere, F. (2001). 'Pulsed light for food decontamination: A review'. *Trends in Food Science and Technology*, *18*, 464–473.
63. Granato, D., Branco, G. F., Nazzaro, F., Cruz, A. G., & Faria, J. A. F. (2010). Functional foods and nondairy probiotic food development: Trends, concepts, and products. *Comprehensive Reviews in Food Science and Food Safety*, *9*, 292–302.
64. Gryn-Rynko, A., Bazylak, G., & Olszewska-Slonina, D. (2016). New potential phytotherapeutics obtained from white mulberry (*Morus alba* L.) leaves. *Biomedicine and Pharmacotherapy, Elsevier Masson SAS*, *84*, 628–636.
65. Gulati, K., Rai, N., Chaudhary, S., & Ray, A. (2016). Nutraceuticals in respiratory disorders. In, R. C. Gupta (Ed.), *Nutraceutical: Efficacy, Safety and Toxicity*. Academic Press.
66. Gunja, N., & Brown, J. A. (2012). Energy drinks: Health risks and toxicity. *The Medical Journal of Australia*, *196*, 46–49.
67. Hasler, C. M. (2008). Health claims in the United States: An aid to the public or a source of confusion?. *The Journal of Nutrition*, *138*(6), 1216S–1220S.

68. Heckman, M. A., Sherry, K., & Gonzalez de, M. E. (2010). Energy drinks: An assessment of their market size, consumer demographics, ingredient profile, functionality, and regulations in the United States. *Comprehensive Reviews in Food Science and Food Safety*, *9*, 303–317.
69. Hendrickx, M. E. G., & Knorr, D. W. (2002). *Ultra High-Pressure Treatments of Foods*. Kluwer Academic/Plenum Publishers.
70. Hmid, I., Elothmani, D., Hanine, H., Oukabli, A., & Mehinagic, E. (2017). Comparative study of phenolic compounds and their antioxidant attributes of eighteen pomegranates (*Punica granatum* L.) cultivars grown in Morocco. *Arabian Journal of Chemistry*, *10*(2), 2675–2684.
71. Hoefler, A. C. (2004). *Practical Guides for Food Industry, Hydrocolloids, Eagan Press Handbook Series* (pp. 33–34).
72. Huang, H. W., Wu, S. J., Lu, J. K., Shyu, Y. T., & Wang, C. Y. (2017). Current status and future trends of high-pressure processing in the food industry. *Food Control*, *72*, 1–8.
73. Hruškar, M., Levaj, B., Kovačević, D. B., Kićanović, S. (2012). Sensory profile of plum nectars. *Croatian Journal of Food Technology, Biotechnology and Nutrition*, *7*(Special Issue), 28–33.
74. Hwang, S. H., & Hong, J. H. (2015). Determining the most influential sensory attributes of nuttiness in soymilk: A trial with Korean consumers using model soymilk systems. *Journal of Sensory Studies*, *30*(5), 425–437.
75. Jackson, C. J. C., & Paliyath, G. (2011). Functional foods, and nutraceuticals. In G. Paliyath,., M. Bakovic., & K. Shetty (Eds.), *Functional Foods, Nutraceuticals, and Degenerative Disease Prevention* (pp. 11–43). Oxford, UK: Wiley-Blackwell.
76. Jae, K. K., Shin, E. C., Lim, H. J., Choi, S. J., Kim, C. R., Suh, S. H., Kim, C. J., Park, G. G., Park, C. S., Kim, H. K., Choi, J. H., Song, S. W., & Shin, D. H. (2015). Characterization of nutritional composition, antioxidative capacity, and sensory attributes of Seomae mugwort, a native Korean variety of Artemisia argyi H. Lév. & Vaniot. *Journal of Analytical Methods in Chemistry*, 1–9.
77. Jambrak, A. R., Šimunek, M., Petrović, M., Bedić, H., Herceg, Z., & Juretić, H. (2017). Aromatic profile and sensory characterization of ultrasound treated cranberry juice and nectar. *Ultrasonics Sonochemistry*, *38*, 783–793.
78. Jaworska, G., Grega, T., Sady, M., Bernaś, E., & Pogoń, K. (2014). Quality of apple-whey and apple beverages over the 12-month storage period. *Journal of Food and Nutrition Research*, *53*(2), 117–126.
79. Kalman, D. S., Feldman, S., Krieger, D. R., & Bloomer, R. J. (2012). Comparison of coconut water and a carbohydrate-electrolyte sport drink on measures of hydration and physical performance in exercise-trained men. *Journal of the International Society of Sports Nutrition*, *9*(1), 1.
80. Jiang, Y., & Nie, W. J. (2015). Chemical properties in fruits of mulberry species from the Xinjiang province of China. *Food Chemistry, Elsevier Ltd*, *174*, 460–466.
81. Kawashima, L. M., & Valente Soares, L. M. (2003). Mineral profile of raw and cooked leafy vegetables consumed in Southern Brazil. *Journal of Food Composition and Analysis*, *16*(5), 605–611.
82. Khandpur, P., & Gogate, P. R. (2015). Understanding the effect of novel approaches based on the ultrasound on the sensory profile of orange juice. *Ultrasonics Sonochemistry*, *27*, 87–95.

83. Kim, H. J., Kim, P., & Shin, C. Y. (2013). A comprehensive review of the therapeutic and pharmacological effects of ginseng and ginsenosides in the central nervous system. *Journal of Ginseng Research, 37*(1), 8–29.
84. Kinsella, J. E., & Damodaran, S. (1980). Flavor problems in soy proteins: Origin nature, control and binding phenomena. In G. Charalambous (Ed.), *The Analysis and Control of Less Desirable Flavors in Foods and Beverages* (pp. 95–109). New York: Academic Press.
85. Knorr, D. (1993). Effect of high hydrostatic pressure processes on food safety and quality. *Food Technology, 47*, 156–161.
86. Knorr, D. (2002). High-pressure processing for preservation, modification, and transformation of foods. *High-Pressure Research, 22*(3–4), 595–599.
87. Knorr, D., Heinz, V., & Buckow, R. (2006). High-pressure application for food biopolymers. *Biochimica et Biophysica Acta, 1764*(3), 619–631.
88. Kocadağli, T., Özdemir, K. S., & Gökmen, V. (2013). Effects of infusion conditions and decaffeination on free amino acid profiles of green and black tea. *Food Research International, 53*(2), 720–725.
89. Kratchanova, M., Pavlova, E., & Panchev, I. (2004). The effect of microwave heating of fresh orange peels on the fruit tissue and quality of extracted pectin. *Carbohydrate Polymers, 56*, 181–186.
90. Kumar, R. S. (2015). Development, quality evaluation and shelf life studies of probiotic beverages using Whey and Aloe vera Juice. *Journal of Food Processing and Technology, 06*(09).
91. Kumar, B., Vijayendra, S. V. N., & Reddy, O. V. S. (2015). Trends in dairy and non-dairy probiotic products–a review. *Journal of Food Science and Technology, 52*(10), 6112–6124.
92. Lal, G. G. (2015). Processing of beverages for the health food market consumer. In J. I. Boye (Ed.), *Nutraceutical and Functional Food Processing Technology*. Canada: John Wiley & Sons, Ltd.
93. Lau, T. C., Chan, M. W., Tan, H. P., & Kwek, C. L. (2013). Functional food: a growing trend among the health-conscious. *Asian Social Science, 9*, 198–208.
94. Lawless, L. J. R., Threlfall, R. T., Meullenet, J. F., & Howard, L. R. (2013). Applying a mixture design for consumer optimization of black cherry, concord grape, and pomegranate juice blends. *Journal of Sensory Studies, 28*, 102–112.
95. Lawless, H. T., Sinopoli, D., & Chapman, K. W. (2009). A comparison of the labeled affective magnitude scale and the 9-point hedonic scale and examination of categorical behavior. *Journal of Sensory Studies, 25*, 54–66.
96. Lawrence, N. D., Kentish, S. E., O'Connor, A. J., Barber, A. R., & Stevens, G. W. (2008). Microfiltration of skim milk using polymeric membranes for casein concentrate manufacture. *Separation and Purification Technology, 60*(3), 237–244.
97. Levaj, B., Vahčić, N., Dragović-Uzelac, V., Svetličić, S., Sabljak, V., Herceg, K., Stanić, D., Marinčić, D., Elez, I., Kovačević, D. B., & Lončarić, S. (2012). Influence of processing on yield and quality of cloudy plum juices. *Croatian Journal of Food Technology, Biotechnology and Nutrition, 7*(Special Issue), 34–38.
98. Lee, P. R., Boo, C. X., & Liu, S. Q. (2013). Fermentation of coconut water by probiotic strains *Lactobacillus acidophilus* L10 and *Lactobacillus casei* L26. *Annals of Microbiology, 63*(4), 1441–1450.

99. Liepa, M., Zagorska, J., & Galoburda, R. (2016). High-pressure processing as novel technology in the dairy industry: A review. *Research for Rural Development, 1*(1), 76–83.
100. Lomer, M. C. E., Parkes, G. C., & Sanderson, J. D. (2008). Review article: Lactose intolerance in clinical practice–myths and realities. *Alimentary Pharmacology and Therapeutics, 27*, 93–103.
101. Lopez-Serrano, M., & Barceloä, A. R. (2002). Comparative study of the products of the peroxidase-catalyzed and the polyphenoloxidase-catalyzed (+)-catechin oxidation, their possible implications in strawberry (Fragaria x ananassa) browning reactions. *Journal of Agricultural and Food Chemistry, 50*, 1218–1224.
102. Mana Kialengila, D., Wolfs, K., Bugalama, J., Van Schepdael, A., & Adams, E. (2013). Full evaporation headspace gas chromatography for sensitive determination of high boiling point volatile organic compounds in low boiling matrices. *Journal of Chromatography A, 1315*, 167–175.
103. Markets and Markets. (2018). HPP (High-Pressure Processing) market by equipment type (orientation, vessel size), application (meat, seafood, beverage, fruit and vegetable), product type (meat and poultry, seafood, juice, ready meal, fruit and vegetable) and geography–2013, Forecast to. Report Code: FB 2151.
104. Matta, V. M., Moretti, R. H., & Cabral, L. M. C. (2004). Microfiltration and reverse osmosis for clarification and concentration of acerola juice. *Journal of Food Engineering, 61*(3), 477–482.
105. Mark-Herbert, C. (2004). Innovation of a new product category—Functional foods. *Technovation, 24*, 713–719.
106. Massaro, M., Scoditti, E., Carluccio, M. A., & De Caterina, R. (2010). Nutraceuticals and Prevention of Atherosclerosis: Focus on ω-3 Polyunsaturated Fatty Acids and Mediterranean Diet Polyphenols. *Cardiovascular Therapeutics, 28*, 13–19.
107. McKay, D. L., & Blumberg, J. B. (2002). The role of tea in human health: An update. *Journal of the American College of Nutrition, 21*, 1–13.
108. Mirsaeedghazi, H., Emam-djomeh, Z., & Mohammad, S. (2010). Clarification of pomegranate juice by microfiltration with PVDF membranes. *Desalination, 264*, 243–248.
109. Mitchell, J. B., Grandjean, P. W., Pizza, F. X., Starling, R. D., & Holtz, R. W. (1994). The effect of the volume ingested on rehydration and gastric emptying following exercise-induced dehydration. *Medicine and Science in Sports and Exercise, 26*, 1135–1143.
110. Morell, P., & Fiszman, S. (2017). Revisiting the role of protein-induced satiation and satiety. *Food Hydrocolloids, Elsevier Ltd, 68*, 199–210.
111. Naghma, K., & Hasan, M. (2007). Tea polyphenols for health promotion. *Life Sciences, 81*, 519–533.
112. Naik, L., Sharma, R., Rajput, Y. S., & Manju, G. (2013). Application of high-pressure processing technology for dairy food preservation-future perspective: A review journal of animal production advances. *Journal of Animal Production Advances, 3*(8), 232–241.
113. Nanasombat, S., Thonglong, J., & Jitlakha, J. (2015). Formulation and characterization of novel functional beverages with antioxidant and anti-acetylcholinesterase activities. *Functional Foods in Health and Disease, 5*(1), 1–16.
114. Nicolas, L., Marquilly, C., & O'Mahony, M. (2009). The 9-point hedonic scale: Are words and numbers compatible?. *Food Quality and Preference, 21*, 1008–1015.

115. Niva, M. (2000). Consumers, functional foods, and everyday knowledge. In *Conference of Nutritionists Meet Food Scientists and Technologists*.
116. Oey, I., Van der Plancken, I., Van Loey, A., & Hendrickx, M. (2008). Does high-pressure processing influence nutritional aspects of plant-based food systems? *Trends in Food Science and Technology*, *19*, 300–308.
117. Ogundele, O. M., Awolu, O. O., Badejo, A. A., Nwachukwu, I. D., & Fagbemi, T. N. (2016). Development of functional beverages from blends of Hibiscus sabdariffa extract and selected fruit juices for optimal antioxidant properties. *Food Science and Nutrition*, *4*(5), 679–685.
118. Ohama, H., Ikeda, H., & Moriyama, H. (2006). Health foods and foods with health claims in Japan. *Toxicology*, *22*, 95–111.
119. Oliveira, M., Ramos, S., Delerue-Matos, C., & Morais, S. (2015). Espresso beverages of pure origin coffee: Mineral characterization, contribution for mineral intake, and geographical discrimination. *Food Chemistry*, *177*, 330–338.
120. Omoni, A. O., & Aluko, R. E. (2005). Soybean foods and their benefits: Potential mechanisms of action. *Nutrition Reviews*, *63*, 272–283.
121. Ottaway, P. B. (2009). Fortification of beverages with vitamins and minerals. In P. Paquin (Ed.), *Functional and Specialty Beverage Technology* (pp. 71–91). Cambridge, UK: Woodhead Publishing.
122. Ozen, A. E., Pons, A., & Tur, J. A. (2012). Worldwide consumption of functional foods: A systematic review. *Nutrition Reviews*, *70*, 472–481.
123. Ozer, B. H., & Kirmaci, H. A. (2010). Functional milks and dairy beverages. *International Journal of Dairy Technology*, *63*, 1–15.
124. Oziemblowski, M., & Kopec, W. (2005). Pulsed electric fields (PEF) as an unconventional method of food preservation. *Polish Journal of Food and Nutrition Sciences*, *14*(55), 31–35.
125. Pagliarini, E., Tomaselli, N., & Brenna, O. V. (2004). Study on sensory and composition changes in Italian Amarone valpolicella red wine during aging. *Journal of Sensory Studies*, *19*, 422–432.
126. Paniwynk, L. (2017). Applications of ultrasound in the processing of liquid foods: A review. *Ultrasonics Sonochemistry*, *38*, 794–806.
127. Pereira, C. G., Barreira, L., da Rosa Neng, N., Nogueira, J. M. F., Marques, C., Santos, T. F., Varela, J., & Custódio, L. (2017). Searching for new sources of innovative products for the food industry within halophyte aromatic plants: In vitro antioxidant activity and phenolic and mineral contents of infusions and decoctions of *Crithmum maritimum* L. *Food and Chemical Toxicology*, *107*, 581–589.
128. Pinto, E., Almeida, A. A., Aguiar, A. A. R. M., & Ferreira, I. M. P. L. V. O. (2014). Changes in macrominerals, trace elements, and pigments content during lettuce (Lactuca sativa L.) growth: Influence of soil composition. *Food Chemistry*, *152*, 603–611.
129. Prado, F. C., Parada, J. L., Pandey, A., & Soccol, C. R. (2008). Trends in non-dairy probiotic beverages. *Food Research International*, *41*, 111–123.
130. Pravst, I. (2012). Functional foods in Europe: A focus on health claims. In B. Valdez (Ed.), *Scientific, Health, and Social Aspects of the Food Industry* (pp. 165–208). Rijeka, Croatia: InTech.
131. Chandra, C. T., & Rao, P. S. (2013). Shelf-life and color change kinetics of Aloe vera gel powder under accelerated storage in three different packaging materials. *Journal of Food Science and Technology*, *50*(4), 747–754.

132. Ravikumar, C. (2014). Review on herbal teas. *Journal of Pharmaceutical Sciences and Research, 6*(5), 236–238.
133. Rifnaz, M. B. M., Jayasinghe-Mudalige, U. K., Guruge, T. P. S. R., Udugama, J. M. M., Herath, H. M. L. K., & Edirisinghe, J. C. (2016). Perceived health status of consumers and incorporation of functional ingredients into their diet. *Procedia Food Science, 6*, 56–59.
134. Ripps, H., & Shen, W. (2012). Review–taurine: A "very essential amino" acid. *Molecular Vision, 18*, 2673–2686.
135. Rodriguez-Fragoso, L., Reyes-Esparza, J., Burchiel, S., Herrera-Ruiz, D., & Torres, E. (2008). Risks and benefits of commonly used herbal medicines in Mexico. *Toxicology and Applied Pharmacology, 227*(1), 125–135.
136. Rodriguez-Roque, M. J., Rojas-Grau, M. A., Elez-Martinez, P., & Martin-Belloso, O. (2013). Soymilk phenolic compounds, isoflavones and antioxidant activity as affected by in vitro gastrointestinal digestion. *Food Chemistry, 136*, 206–212.
137. Rossi, M., Giussani, E., Morelli, R., Lo Scalzo, R., Nani, R. C., & Torregiani, D. (2003). Effect of fruit blanching on phenolics and radical scavenging activity of highbush blueberry juice. *Food Research International, 36*, 999–1005.
138. Rupasinghe, H. P. V., & Yu, L. J. (2012). Prof. Yehia El-Samragy (Ed.), *Emerging Preservation Methods for Fruit Juices and Beverages, Food Additive*. InTech.
139. Rysstad, G., & Johnstone, K. (2009). Extended shelf-life beverages. In P. Paquin (Ed.), *Functional and Specialty Beverage Technology* (pp. 107–132).
140. Saarela, M. (2009). Probiotics as ingredients in functional beverages. In P. Paquin (Ed.), *Functional and Speciality Beverage Technology* (pp. 55–70).
141. Sala-Vila, A., Estruch, R., & Ros, E. (2015). New insights into the role of nutrition in CVD prevention. *Current Cardiology Reports, 17*(5), 26.
142. Salvia-Trujillo, L., Morales-de la Peña, M., Rojas-Graü, A., Welti-Chanes, J., & Martín-Belloso, O. (2017). Mineral and fatty acid profile of high intensity pulsed electric fields or thermally treated fruit juice-milk beverages stored under refrigeration. *Food Control, 80*, 236–243.
143. Sampedro, L., Moreira, X., Martíns, P., & Zas, R. (2009). Growth and nutritional response of *Pinuspinaster* after a large pine weevil (*Hylobiusabietis*) attack. *Trees, 23*(6), 1189–1197.
144. Sanchis-Gomar, F., Pareja-Galeano, H., Cervellin, G., Lippi, G., & Earnest, C. P. (2015). "Energy drink overconsumption in adolescents: Implications for arrhythmias and other cardiovascular events." *Canadian Journal of Cardiology, 31*(5), 572–575.
145. Servili, M., Rizzello, C. G., Taticchi, A., Esposto, S., Urbani, S., Mazzacane, F., Di Maio, I., Selvaggini, R., Gobbeti, M., & Di Cagno, R. (2011). Functional milk beverage fortified with phenolic compounds extracted from olive vegetation water and fermented with functional lactic acid bacteria. *International Journal of Food Microbiology, 147*(1), 45–52.
146. Sfakianakis, P., & Tzia, C. (2014). Conventional and Innovative processing of milk for yogurt manufacture; development of texture and flavor: A review. *Foods, 3*(1), 176–193.
147. Sharma, S., Vaidya, D., & Rana, N. (2016). Honey as a natural sweetener in the lemon ready-to-serve drink. *International Journal of Bio-resource and Stress Management, 7*(2), 320–325.
148. Siguemoto, É. S., Gut, J. A. W., Martinez, A., & Rodrigo, D. (2018). Inactivation kinetics of Escherichia coli O157: H7 and Listeria monocytogenes in apple juice

by microwave and conventional thermal processing. *Innovative Food Science and Emerging Technologies*, *45*, 84–91.
149. Srivastava, J. K., Shankar, E., & Gupta, S. (2010). Chamomile: A herbal medicine of the past with a bright future. *Molecular Medicine Reports*, *3*(6), 895–901.
150. Stratakos, A. C., Delgado-Pando, G., Linton, M., Patterson, M. F., & Koidis, A. (2016). Industrial-scale microwave processing of tomato juice using a novel continuous microwave oven. *Food Chemistry*, *190*, 622–628.
151. Stroup, B. M., Murali, S. G., Nair, N., Sawin, E. A., Rohr, F., Levy, H. L., & Ney, D. M. (2017). Dietary amino acid intakes associated with a low-phenylalanine diet combined with amino acid medical foods and glycomacropeptide medical foods and neuropsychological outcomes in subjects with phenylketonuria. *Data in Brief*, *13*, 377–384.
152. Sun-Waterhouse, D. (2011). The development of fruit-based functional foods targeting the health and wellness market: A review. *International Journal of Food Science and Technology*, *46*, 899–920.
153. Sushilkumar, S. M., Sawate, A. R., Patil, B. M., Kshirsagar, R. B., & Kulkarni, S. P. (2016). Study on the effect of custard apple leaf extract on physico-chemical properties of aonla juice. *Asian Journal of Dairy and Food Research*, *35*(1), 81–84.
154. Suzuki, Y., Miyoshi, N., & Isemura, M. (2012). Health-promoting effects of green tea. *Proceedings of the Japan Academy. Series B, Physical and Biological Sciences*, *88*(3), 88–101.
155. Švarc-Gajić, J., Cvetanović, A., Segura-Carretero, A., Mašković, P., & Jakšić, A. (2017). Functional coffee substitute prepared from ginger by subcritical water. *The Journal of Supercritical Fluids*, *128*, 32–38.
156. Sworn, G. (2007). Introduction to food hydrocolloids. In P. A. Williams (Ed.), *Handbook of Industrial Water Soluble Polymers* (Vol. 18, p. 24). Cambridge: Blackwell Publishing.
157. Szilagyi, A. (2015). Adaptation to lactose in lactase non-persistent people: Effects on intolerance and the relationship between dairy food consumption and evaluation of diseases. *Nutrients*, *7*(8), 6751–6779.
158. Tripathi, A. K., & Misra, A. K. (2005). Soybean–A consummate functional food: A review. *Journal of Food Science and Technology*, *2*, 42–46.
159. Turk, M., Perino, S., Cendres, A., Petitcolas, E., Soubrat, T., & Chemat, F. (2017). Alternative process for strawberry juice processing : Microwave hydrodiffusion and gravity. *LWT-Food Science and Technology*, *84*, 626–633.
160. Udabage, P., Augustin, M., Versteeg, C., Puvanenthiran, A., Yoo, J., Allen, N., & Kelly, A. (2010). Properties of low-fat stirred yogurts made from high-pressure-processed skim milk. *Innovative Food Science and Emerging Technologies*, *11*(1), 32–38.
161. Vaillant, F., Cisse, M., Chaverri, M., Perez, A., Dornier, M., Viquez, F., & Dhuique-Mayer, C. (2005). Clarification and concentration of melon juice using membrane processes. *Innovative Food Science and Emerging Technologies*, *6*(2), 213–220.
162. Vieira da Silva, B., Barreira, J. C. M., & Oliveira, M. B. P. P. (2016). Natural phytochemicals and probiotics as bioactive ingredients for functional foods: Extraction, biochemistry and protected-delivery technologies. *Trends in Food Science and Technology*, *50*, 144–158.
163. Venkatesh, M., & Raghavan, G. S. V. (2004). An overview of microwave processing and dielectric properties of agri-food materials. *Biosystems Engineering*, *88*(1), 1–18.

164. Verma, A., & Singh, A. (2013). Optimization of ingredients for a herbal beverage with medicinal attributes. *Asian Journal of Dairy and Food Research, 32*(4), 318–322.
165. Wang, L., & Weller, C. L. (2006). Recent advances in extraction of nutraceuticals from plants. *Trends in Food Science and Technology, 17*, 300–312.
166. Wang, B., Kammer, L. M., Ding, Z., Lassiter, D. G., Hwang, J., Nelson, J. L., & Ivy, J. L. (2012). Amino acid mixture acutely improves the glucose tolerance of healthy overweight adults. *Nutrition Research, 32*(1), 30–38.
167. Warriner, K., Movahedi, S., & Waites, W. M. (2004). 'Laser-based packaging sterilization in aseptic packaging'. In P. Richardson (Ed.), *Improving the Thermal Processing of Foods*. Cambridge: Woodhead Publishing.
168. Wolf, P. H., Siverns, S., & Monti, S. (2005). UF membranes for RO desalination pretreatment. *Desalination, 182*(1e2), 293–300.
169. Xing Chen, L., Xing, Y., Cao, L., Xu, Q., Li, S., Wang, R., Jiang, Z., Che, Z., & Lin, H. (2017). Effects of six commercial Saccharomyces cerevisiae strains on phenolic attributes. Antioxidant activity, and aroma of kiwifruit (*Actinidiadeliciosacv.*) wine. *BioMed Research International*, 1–10. Article ID 2934743
170. Yang, Y. (2008). Scientific substantiation of functional food health claims in China. *Journal of Nutrition, 138*, 1199S–205S.
171. Yuan, L., Xiong, L., Deng, T., Wu, Y., Li, J., Liu, S., Huang, J., & Liu, Z. (2015). Comparative profiling of gene expression in *Camellia sinensis* L. cultivar AnJiBaiCha leaves during periodic albinism. *Gene, 561*(1), 23–29.
172. Yuan, T., Li, L., Zhang, Y., & Seeram, N. P. (2013). Pasteurized and sterilized maple sap as functional beverages: Chemical composition and antioxidant activities. *Journal of Functional Foods, 5*(4), 1582–1590.
173. Zhao, Y., Flugstad, B., Kolbe, E., Park, J. W., & Wells, J. H. (2000). Using capacitive (radio frequency) dielectric heating in food processing and preservation-a review. *Journal of Food Process Engineering, 23*(1), 25–55.
174. Anonymous. (2012). How the food industry defines botanicals (2012, June 05). Retrieved from https://www.foodprocessing.com/articles/2012/defining-botanicals/ (accessed March 20, 2019).
175. Beer, E. (2017). Storming growth for energy, water, and tea in functional beverages. Retrieved from https://www.nutraingredients.com/Article/2017/03/15/Storming-growth-for-energy-water-and-tea-in-functional-beverages (accessed March 13, 2019).
176. Cowland, D. (2013). Health and wellness beverages outperform the wider drinks industry. Retrieved from http://blog.euromonitor.com/2013/02/health-and-wellness-beverages-outperform-wider-drinks-industry.html (accessed March 13, 2019).
177. Diabetes.co.uk. (2017). Sports drinks. Retrieved from http://www.diabetes.co.uk/sports-drinks.html (accessed March 20, 2019).
178. Enos, D. (2011). 'Superfruits' may bring some health benefits. (November 10, 2011). Retrieved from https://www.livescience.com/35968-superfruit-pomegranate-acai-gogi-health-benefits.html (accessed March 20, 2019).
179. Euromonitor. (2015). Global state of the industry: A review of the top tea trends and markets around the world. Retrieved from https://www.slideshare.net/Euromonitor/euromonitor-world-tea-expo-final (accessed March 20, 2019).
180. Euromonitor. (2017a). RTD tea in Malaysia. Retrieved from http://www.euromonitor.com/rtd-tea-in-malaysia/report (accessed March 20, 2019).

181. Euromonitor. (2017b). RTD tea in Australia. Retrieved from http://www.euromonitor.com/rtd-tea-in-australia/report (accessed March 20, 2019).
182. FAO/WHO. (2006). Probiotics in food: Health and nutritional properties and guidelines for evaluation. Retrieved from http://www.fao.org/3/a-a0512e.pdf (accessed March 13, 2019).
183. Food Insight. (2009). Functional foods fact sheet: Soy. (October 14, 2009). Retrieved from http://www.foodinsight.org/Functional_Foods_Fact_Sheet_Soy (accessed March 13, 2019).
184. Fortitech. (2011). New innovation in functional beverages. Retrieved from https://www.fortitechpremixes.com/wp.../Functional_Beverages_FINAL_ENG.pdf (accessed March 14, 2019).
185. Global Industry Analysts, Inc. (2016). Innovative flavors and organic & functional juices to sustain growth in the global fruit and vegetable juices market. Retrieved from http://www.strategyr.com/MarketResearch/Fruit_and_Vegetable_Juices_Market_Trends.asp (accessed March 14, 2019).
186. Kubitz, L. (2016, April 07). Super Berry: Is Acai a 'Superfood'?. Retrieved from http://www.foodinsight.org/acai-superfood-antioxidants-flavonoids-blueberries (accessed March 20, 2019).
187. Statisca. (2017). Sales volume of soy milk worldwide in 2015 and 2018 (in billion liters). Retrieved from https://www.statista.com/statistics/645662/soy-milk-sales-volume-worldwide/ (accessed March 20, 2019).

CHAPTER 12

Oestradiol, Aging Theory, Women/LGBT Difficulties and Motivation

FRANCISCO TORRENS[1] and GLORIA CASTELLANO[2]

[1]Institut Universitari de Ciència Molecular, Universitat de València, Edifici d'Instituts de Paterna, P.O. Box 22085, E-46071 València, Spain, E-mail: torrens@uv.es

[2]Departamento de Ciencias Experimentales y Matemáticas, Facultad de Veterinaria y Ciencias Experimentales, Universidad Católica de Valencia San Vicente Mártir, Guillem de Castro-94, E-46001 València, Spain

ABSTRACT

An interface is more than a geometrical boundary. No scientific evidence exists that the supplements of antioxidants reduce the risk of degenerative diseases and lengthen life. Hypotheses on the reasons for the apparent lack of efficiency follow: compounds resulting from redox reactions are not all chemically stable and do not prevent the progression of the chain oxidizing reaction; when administered excessively, they could interfere with vital oxidizing reactions, which are not so bad, they serve as signs of communication and activation of natural defenses, immunity or reparative that avoid worse paths. Although people take amounts of antioxidants, their presence in the tissues does not show a direct relation to the ingestion, and it seems that the attributed antioxidant effect could be done not only by direct action but inducing the synthesis of other organic antioxidants *via* the expression of genes encoding them. It is healthy and appropriate to obtain a natural diet, rich in antioxidants; but it does not seem reasonable to recommend using commercial preparations in megadosis. Some of the most powerful natural antioxidants are thiols and polyphenols, which make up numerous families. In order to study, one must be motivated.

12.1 INTRODUCTION

Setting the scene: estradiol, the main female hormone, theories of aging, free radicals (FRs), antioxidants (AOs), biosensors (BSs), in search for Techniques 4.0, the women of the future, the contributions and difficulties of women and lesbian gay bisexual transgender, etc. (LGBT+) people in science, reflections on a controversy, students' motivation towards chemistry, and strategies for job search. Perhaps, the first theory of aging was extracted from the environment by a simple observation: Things wear out naturally by use and by the passage of time, but, why do some things or people last longer than others? The future is getting closer to every day. Moreover, it seems that the future will be marked in many ways by the digital revolution in which people are immersed, which does not lose strength. If people are exploring and adapting themselves to the consequences of the arrival of the Internet and the information society, they see the ears of challenges (e.g., artificial intelligence, automation of sites of work), that is, the replacement of human personnel for machines in increasingly complex tasks. The global discussion on the issues began, and it is interesting to follow the debates about their social and economic consequences (*cf.* Figure 12.1).

The motivation of the students to the learning of any scientific matter constitutes a serious preoccupation for the teaching staff [1]. First, because teachers show a general lack of interest in students and, above all, to physics and chemistry, which perception is corroborated by the students. Anecdotally, a Spanish Ministry of Education and Science survey about the school failure of secondary-education students showed that 25% said they were not interested in scientific studies, and 68% indicated that they were bored in class. Second, because next to the lack of motivation, a school failure exists in the disciplines greater than in others. Students' disinterest rises as secondary-school students receive more courses in the subjects. The teachers meet the following *vicious circle*: Students come to the chemistry class demotivated; this means that they do not pay attention to the teacher's explanations; as a result, they get bored, disinterest in learning rises and they do not make any effort to learn. How to break the *disinterest-poor academic performance-greater disinterest* spiral?

In earlier publications, it was informed ultraviolet-visible (UV-VIS) *absorption* spectroscopy, its link with *fluorescence* (FLU) [2], FLU *excitation*, *emission*, its comparison with UV-VIS absorption [3] and *mass spectrometry*, the only method with many ionization techniques [4].

FIGURE 12.1 (a) Programmers Jean Jennings (left), Marlyn Wescoff (center) and Ruth Lichterman working on the ENIAC computer *ca.* 1946, (b) Grace Hopper, inventor of the first compiler and programming language COBOL, (c) Margaret Hamilton in a simulator of command module of Apollo missions.

It was reported metaphors that made history, reflections on philosophy, science, deoxyribonucleic acid [5], the proposal of a journal of science-technology and gender spreading [6], research with a gender perspective, also in experimental science [7], service-learning, employability, gender diversity, a gap in science-technology-engineering-mathematics (STEM) [8], computation, quantum mechanics, metaphors, feminist tools [9], metal-organic frameworks, science, pseudo-science, chronobiology and getting women wrong [10]. The purpose of the present report is to review estradiol, the main female hormone, theories of aging, FRs, AOs, BSs, in search for Techniques 4.0, the women of the future, the contributions and difficulties of women and LGBT people in science, reflections on a controversy, students' motivation towards chemistry and strategies for job search. The aim of this work is to initiate a debate by suggesting a number of questions (Q), which can arise when addressing subjects of hormones, aging, FRs, AOs, BSs, the women of the future, the contributions and difficulties of women and LGBT people in science, students' motivation towards chemistry and strategies for job search. It was provided, when possible, answers (A) and hypotheses (H) on these materials, processes, and ideas.

12.2 ESTRADIOL: THE MAIN FEMALE HORMONE

Estrogen (OEST) is the name given to a group of compounds, for example, estradiol (*cf.* Figure 12.2), which are important in the female reproductive cycle [11]. The name comes from the Greek *oistros* meaning *producer of sexual desire*. Like the male equivalent, testosterone, OESTs are produced in all vertebrates and insects, which suggests that they appeared early on in the evolution of life on Earth. It has to do with sex in both gender and reproductive meaning of the word. The sex hormones determine whether an animal is male or female, and the relative amounts of the different sex hormones decide the sexual characteristics of that animal, for example, how masculine or feminine they appear and behave.

The female sex hormone estradiol is the most important one, along with estrone and estriol, which are less potent (*cf.* Figure 12.3) [11]. They are all based on the standard four-ring steroid structure (*four rooms and a bathroom*), with just minor differences inside groups, which includes testosterone. It is quite amazing that all the differences between males and females come down to something as trivial as the side group of a molecule!

Although it can be produced by many cells in the body [e.g., fat cells, in the brain, artery walls (in males and females)], most estradiol is synthesized in the ovaries from compounds derived from cholesterol (CHOL). Estradiol helps a woman prepare for pregnancy, supporting the reproductive organs, keeping the eggs healthy in the ovaries, and instigating the monthly ovulation and menstrual cycle. It is responsible for the development of the female secondary sex characteristics, which begin at puberty and decline after the menopause; for example, estradiol initiates the development of breasts and alters the fat distribution in a woman's body to make her curvier. It helps strengthen bones and joints.

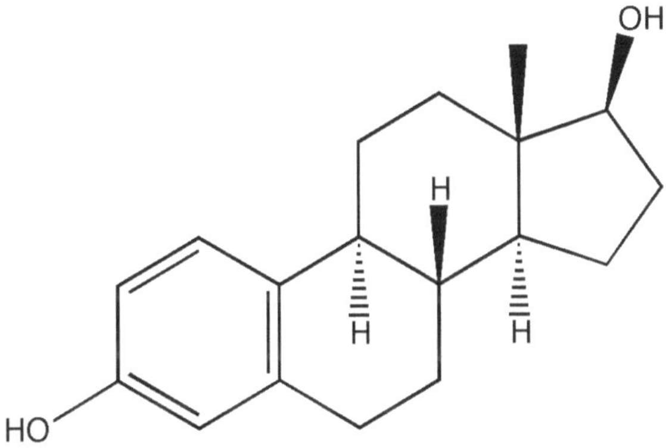

FIGURE 12.2 Molecular structure of estradiol.

Estradiol is partially why post-menopausal women present brittle bones [11]. After menopause, women's OEST levels drop dramatically, which results in their bones becoming weak and porous (*osteoporosis*). To combat this, post-menopausal women, or those who had a hysterectomy, are usually prescribed hormone replacement therapy (HRT), which typically involves ingesting a mixture of OEST, progesterone, and progestin to compensate for the amount they lost. One common OEST that is used in HRT is *pregnant mares' urine* (*Premarin*), which is isolated from the urine of pregnant mares. It is actually mostly composed of estrone sulfate (*cf.* Figure 12.4). It may sound an odd thing to take, but this is converted directly into normal estradiol in the woman's body.

Exposure to any of the components of OEST by males can cause anorexia, vomiting, and feminization, for example, breast growth, and erectile dysfunction [11]. In the developed world, sperm counts fell 1–2% every year for the past several decades. The reason for this is not entirely clear, but one possibility is that men are inadvertently ingesting small quantities of OESTs or estrogenic-like compounds, which are damaging male fertility. A number of chemicals were blamed for possibly mimicking the effects of OEST, for example, bisphenol A (BPA), polychlorinated biphenyls (PCBs), and phthalates (*cf.* Figure 12.5). However, another route *via* which OESTs made their way into H_2O supply is a result of the disposal of unwanted or out-of-date steroid and contraceptive pills down the sink or toilet.

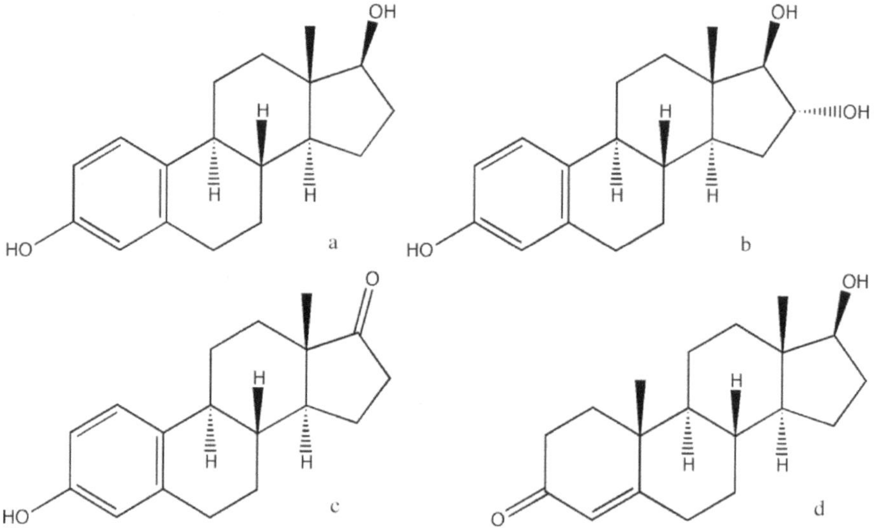

FIGURE 12.3 (a) Estradiol, 2 OH groups, hence the-*diol* suffix, (b) estriol, 3 OH groups, hence -*triol*, (c) Estrone, 1 OH group, and 1 ketone group, hence -*one*, (d) testosterone (the male hormone); note the difference with estradiol is only an extra CH_3 group, the OH was oxidized to a ketone and the phenyl ring, reduced.

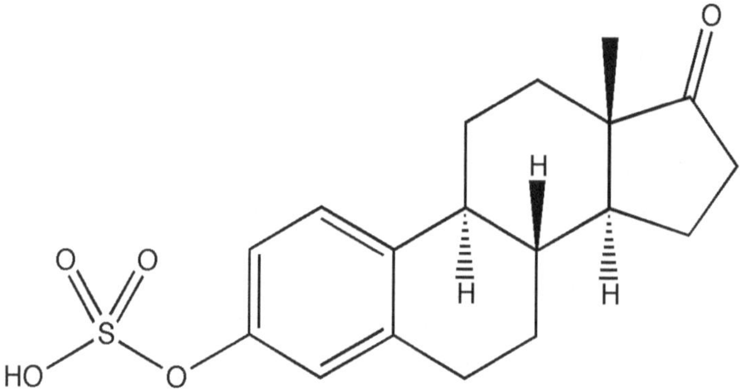

FIGURE 12.4 Molecular structure of estrone sulfate.

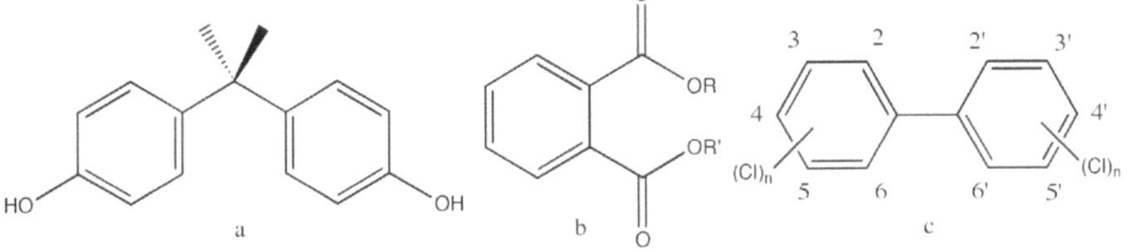

FIGURE 12.5 (a) BPA is used to make polycarbonate plastic and the epoxy resins that are used as a lining in most food and beverage cans, (b) phthalates are plasticizers providing durability and flexibility to plastics, for example, poly(vinyl chloride) (PVC); high-molecular-weight (MW) phthalates are used in wall coverings, flooring, and intravenous bags and tubing; low-MW phthalates are found in cosmetics (e.g., lotions, perfumes), varnishes and lacquers, (c) PCBs are chemically stable, have low flammability, are electrically insulating, and their main use is as insulating fluids and coolants.

The female contraceptive pill contains the hormone progesterone (*the pregnancy hormone*) combined with an OEST, usually a derivative of estradiol (ethynyl estradiol, *cf.* Figure 12.6) [11]. High levels of progesterone, for example, those in the pill, are usually only found when a woman is pregnant, so the pill works by tricking the body into thinking it is already pregnant. The female body's response to being pregnant is complex. However, among other effects, further ovulation is prevented (because no more eggs are needed if one was already fertilized), the mucus in the cervix thickens making it difficult for additional sperm to enter, and the lining of the womb becomes thinner, so it is less likely to accept (another) fertilized egg. Other contraceptive pills, for example, the progesterone-only pill (mini-pill), contain only synthetic progestogens (compounds chemically related to progesterone but not this) and no OEST. Nevertheless, all of the pills contain compounds that mimic OEST, which can present hormone-disruptive effects on wildlife and humans.

People should be worried [11]. The male-to-female ratio in many animals and fish species was altered by exposure to the chemicals. In some of the cases, the males were *feminized* and present partial female characteristics. In addition, in humans, they are thought to be responsible for premature puberty in regions where high levels were found in the local environment. Several countries already started to ban the estrogenic compounds; for example, Canada, the US, and the EU banned (2012) the use of BPA in baby bottles. The PCBs were banned in 1979 but do not degrade effectively, so continue to persist in the environment. However, many more natural and man-made estrogenic compounds exist all-around people, the effects of which they know little or nothing about. Maybe everyone should all start to get used to wearing skirts and dresses!

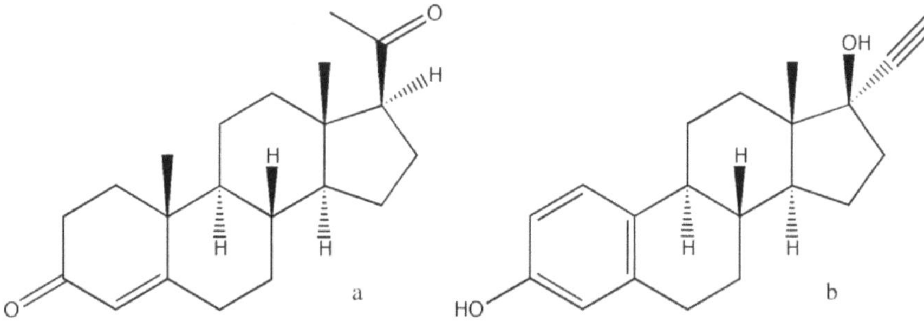

FIGURE 12.6 (a) Progesterone (*the pregnancy hormone*); note the remarkable structural similarity to testosterone (Figure 3d), (b) ethynyl estradiol (the main difference between this and estradiol is the addition of the acetylene group).

12.3 THEORIES OF AGING: FREE RADICALS AND ANTIOXIDANTS

In life chemistry, necessary metabolic substances exist that are harmful or toxic when presented at inadequate levels [12]. Assuming that no such aggressive genetic defect exists, the environment plays an important role as an accelerator, brake, or trigger, since in the metabolic processes of life, biochemical damage and errors slowly accumulate in the macromolecules that end to be irreparable from the moment they alter cellular functions. After more than twenty years of study, the effectiveness results show doubts: In most cases, the benefits showed to be neutral or harmless, and, in some cases, showed deleterious effects.

12.4 BIOSENSORS: IN SEARCH FOR TECHNIQUES 4.0

Tortajada Genaro proposed questions, answers, and hypotheses on BS in search of techniques 4.0 [13].

Q1. What is Industry 4.0?
Q2. What is the role of chemistry?
Q3. What analytical techniques (ATs) are compatible with the integrated systems (ISs) paradigm?
Q4. What is a BS?

A4. A BS is a sensor in a *bio*-context.
H1. Elements of a BS: bioreceptor (BR); support; transducer.
H2. BR: It is not enough that the molecules remain inside the cell.
H3. It is not enough that the molecules remain in the water solvent.
H4. These molecules could not be the same *in vivo* as *in vitro*.
H5. Specificity: complex; affinity elements; nucleic acids (NAs).
H6. Complex: enzyme-substrate complex; the whole cell as BS system.
H7. Affinity elements: complex formation reactions, for example, antigen-antibody.
H8. NAs: A–T and C–G sequence recognitions.
Q5. Can chemists copy this?
A5. Yes, with biomimetic molecules, for example, molecularly imprinted polymer (MIP)-enzyme complex.
H9. Support: materials.
Q6. How does one anchor those BRs on a surface?
A6. Playing with weak forces; entrapment; cross-links; generating bonds in BRs.
H10. Playing with weak forces: adsorption.
H11. Entrapment: in a polymeric net.
H12. Cross-links: generating polymers from BRs.
H13. Generating bonds in BRs: click chemistry.
H14. Transducer: group with redox characteristics anchored to the electrode; optical; piezoelectric.
H15. Optical transducer: surface plasmon resonance (SPR).
H16. Piezoelectric transducer: piezoelectric change.
H17. For example: glucometer [recognizes glucose *via* glucose oxidase (GOx) enzyme, electrochemical]; pregnancy test; acquired immunodeficiency syndrome (AIDS)/malaria/dengue/Zika diagnoses; disease biomarkers (BMs, cancer); food control.
H18. Applications: many, related to daily life; for diagnosis (disease/health)/defense/food safety.
H19: Future perspectives: monitored houses; watches; fabrics.
Q7. Do the BRs people currently have to serve?
A7. Yes.
Q8. Can people improve them?
A8. Potentially, yes.
H20. Challenges: new materials; transduction; formats; multifunction; miniaturization; reuse; cost.

H21. Transduction: new modes of electromagnetic radiation (EMR)-matter interaction.
H22. New formats: Classical chemistry takes many steps, for example, a toxic compound in a meat product.
H23. Multifunction: a glucometer does not do more things.
H24. Miniaturization is better.
H25. Cost reduction.
H26. Future: nanotechnology; integration of ATs.
H27. Nanotechnology: substitute C nanotubes, graphene (GR), reduced GR oxide for BR.
H28. Integration of ATs:
Classical chemistry ® Instrumental techniques ® Robotics ® Nano/Microsystems
H29: Advantages of (all-in-a-chip) ISs: primary and secondary ones.
H30: Primary advantages: lower volume; lower sample volume; less waste.
H31: Secondary advantages: greater portability; greater sensitivity; faster.
Q9. Do all clients benefit?
A9. No, because they are large equipment, require specialized users, etc.
H32. Improvements: portable equipment; drones (in cases of limited accessibility).
Q10. What do all these systems have in common?
A10. They use miniaturized chips.
H33. Problem: Many are based on classic detection systems, detection has not been overcome.
H34. Other strategies: consumer electronics; conventional use; bioanalytical use.
H35. Conventional use: mobile phone as equipment; electrochemical detectors; medicine.
H36. Medicine: mobile-phone telemedicine; medical insurance with phone communication with the doctor.
H37. Personalized medicine *vs*. patient genomics, for example, a patient who wants to quit smoking but the drug has no effect, sintron/warfarin without effect.
H38. Modified compact disc (CD): an economic technology (CD<1€, CD reader<100€):
25mL blood sample®CD®microfluidic system (60 minutes, no specialized staff, 3 reactants).
H39. For example: testing in the ambulance if one is allergic to penicillin.

H40. Wearables: patches, fabrics, stickers, etc. that monitor the patient oneself.
H41. For example: In athletes, they monitor p*H* in sweat to control hydration and get better performance.
H42. For example: Contact lenses: Taking advantage of tears for BS of pathogens, glucose, infections, etc.
H43. For example: Implants inside the body (internal systems) like glucometer; ingested.
Q11. Multi-analyte analysis, what is the maximum number of analytes that you have analyzed?
A11. Microanalysis: 10 antibiotics, 50 genetic BMs, or 70 toxins in the water.
Q12. How is the whole-cell used?
A12. Approaches: Isolating it; recombinant techniques; whole cell.
H44. Two problems: anchoring and orientation.
Q13. Could it be that more anchoring causes less signal?
A13. Yes, for example, anionic NAs: more anchoring®more electrostatic repulsion®less signal.
Q14. Dou you know what water lilies are?
A14. They are hydrophobic; drag is *via* canals: we do not want maximum anchoring but biorecognition.
Q15. Multidisciplinarity: In the development of a BR, what space is left for the chemist?
A15. Challenge is speaking the same language: Chemists' advantage is that they speak more with non-chemists.

12.5 THE WOMEN OF THE FUTURE

How have people come to the current situation [14]?. Unfortunately, everything changed in the 1980's, in which personal computers, with affordable prices, start trading massively. In the market, competitive, advertising plays a fundamental role. In order to maximize sales, computers are presented as toys, taking advantage of the growing popularity of video games. In addition, since popular ignorance states that all toys present a preferential audience determined by the user's sex, the bright minds of the advertising departments decided to focus on the sale of the male market. The computer ads of the time, in the press and on television, are a sad poem of exacerbated machismo. The stereotype is reinforced in popular culture with films [e.g.,

War Games (1983), *Weird Science* (1985, *cf.* Figure 12.7)] and books [e.g., Steven Levy's *Hackers* (1984)] where all the protagonists are men and women play a secondary role when they are not directly the object of ridicule or sexual denigration. The geek image of the computer expert was established (a young, white, cool, and lonely man) who infiltrated the collective subconscious since then. Women were eliminated from the future.

FIGURE 12.7 The creation of *stereotypes*: (a) *War Games* (1984) frame, (b) *Weird Science* (1985) movie poster. The argument is of rattle : Two high-school students program the *perfect woman*.

12.6 CONTRIBUTIONS AND DIFFICULTIES OF WOMEN AND LGBT PEOPLE IN SCIENCE

Saó-Burjassot organized a Day on *Contributions and Difficulties of Women and LGBT People in Science*. Rueda Pascual proposed H/Q [15].

H1. *Gender gap.*
H2. Vital objective: Women in science-technology-engineering-mathematics (STEM).
H3. The future is STEM-information and communication technology (ICT): People cannot miss it.
H4. *Glass ceiling.*
Q1. Has anyone thought of a woman?

H5. The causes: stereotypes. At six, girls think they are less valid than boys.
H6. Jennifer and John effect: The same CV signed by Jennifer/John, the man gets a higher rating.
H7. *Gender pay gap*.
H8. (Londa Schiebinger). Heart attacks affect women more because symptoms are different and more difficult to detect.
H9. (Londa Schiebinger). Retention system in cars: Dummies are men.
H10. The voice recognition system was tested with men.
H11. Amazon's hiring system has more men, recognizing it as positive and recommending more men.
H12. The consequences: *Penalized* women.
H13. *Scissors plot*: In STEM there are no scissors.
H14. Action lines: social-applicability dissemination; support nets; female-proximity-models visualization.

She provided the following conclusions (Cs).

C1. Having few women in science does not do science so much good.
C2. Women can do science as well as men.
C3. Wasting 50% of people in science is not good for science.

12.7 REFLECTIONS ON A CONTROVERSY: STUDENTS' MOTIVATION TO CHEMISTRY

The main aspects of conventional education that must be criticized in order to promote the motivation of students follow [16].

1. In teaching, the problems that led to the construction of scientific knowledge are not usually presented, without which the introduction of concepts and theories does not make sense.
2. The teaching of chemistry does not show the tentative way scientists plan and try to solve the problems.
3. The science-technology-society (STS) interactions associated with the construction of knowledge are ignored in the teaching of chemistry, in particular.
4. In the teaching of chemistry, the transmission of concepts and theories performed by the teacher predominates, where the essential role of the student is that of the recipient with scarce active participation and without the possibility of regulating their learning.

12.8 STRATEGIES FOR JOB SEARCH

Gosálvez raised some of the most relevant Q that a student should ask before seeking employment [17].

Q1. Why start looking for a job before finishing one's studies at the university?
Q2. What to look for?
Q3. Where to start organizing the job search process?
Q4. When is the best time to start the job search?

12.9 FINAL REMARKS

From the previous results and discussion, the following final remarks can be drawn.

1. An interface is more than a geometrical boundary.
2. No scientific evidence exists that the supplements of antioxidants clearly reduce the risk of degenerative diseases and lengthen the life. The reasons for the apparent lack of efficiency in the real world are not clear, and several hypotheses exist: (a) compounds resulting from redox reactions are not all chemically stable and do not prevent the progression of the chain oxidizing reaction; (b) when administered excessively, they could interfere with vital oxidizing reactions, which are not so bad, they serve as signs of communication and activation of natural defenses, immunity or reparative that avoid worse paths. Although people take important amounts of antioxidants, their presence in the tissues does not show a direct relation to the ingestion, and it seems that the attributed antioxidant effect could be done not only by direct action but also inducing the synthesis of other organic antioxidant substances (e.g., dismutase, catalase, reductase) *via* the expression of genes encoding them.
3. It is healthy and appropriate to obtain a natural diet, rich in antioxidants; but it does not seem reasonable to recommend using commercial preparations in megadosis. Some of the most powerful natural antioxidants are thiols and polyphenols, which make up numerous families (e.g., tannins, flavonoids, carotenoids, lignans, stilbenes), which are abundant in fruits, vegetables, nuts, and seeds. However, the virtues of the natural foods are not only antioxidants, but they are complemented and enhanced by the presence of multivitamins and

fiber, which regulate bowel movements and avoid aggressive proportions of fats and refined sugars. Eating them is highly recommended.

4. Chemistry has a leading role in Industry 4.0. The role of chemists is not replaceable by robots or artificial intelligent systems. Working in a multidisciplinary team: Collaborating, understanding the language of physicians, their calligraphy. Searching for techniques in real-time and close to the point of need. In addition, miniaturizing.
5. An increasing awareness exists in the industry and academic world of the problem of gender stereotypes. Many universities established measures and programs to rebalance the field, and technology companies begin to define the criteria of diversity in the recruitment processes. However, part of the solution is much closer, and people must fight stereotypes every day at home, school, and high school. If people want a future of equality and justice, they must work so that the most important means of production of the economy be not exclusively in the hands of the male half of society. People must recover the women of the future. Having a few women in science does not do science so much good. Women can do science as well as men. Wasting 50% of people in science is not good for science.
6. In order to study, one must be motivated. The motivation is intimately linked to chemistry teaching-learning, problematic situations that are proposed to the students, learning strategies that are developed in the classroom, results in they obtain, the context in which they are learning, etc. All factors contribute to the motivation of the student and achieve the improvement of their processes of inter-regulation or self-regulation, and to achieve a greater autonomy that, counted and debated, is the purpose of any education, for example, chemical.

ACKNOWLEDGMENTS

The authors thanks the support from Fundacion Universidad Catolica de Valencia San Vicente Martir (Project No. 2019-217-001UCV).

KEYWORDS

- **antioxidant**
- **biosensor**
- **female hormone**

- **free radical**
- **LGBT in science**
- **student motivation towards chemistry**
- **woman in science**

REFERENCES

1. Esparza, M., (2016). Ciències o lletres? *Daualdeu, 2016*(10), 16–18.
2. Torrens, F., & Castellano, G., (2020). Ultraviolet-Visible absorption spectroscopy: Link with fluorescence. In: Pourhashemi, A., Deka, S. C., & Haghi, A. K., (eds.), *Research Methods and Applications in Chemical and Biological Engineering* (pp. 37–45). Apple Academic–CRC Press: Waretown, NJ.
3. Torrens, F., & Castellano, G., (2020). Fluorescence excitation and emission: Comparison with absorption. In: Pogliani, L., Torrens, F., & Haghi, A. K., (eds.), *Molecular Chemistry and Biomolecular Engineering: Integrating Theory and Research with Practice* (pp. 51–61). Apple Academic–CRC Press: Waretown, NJ.
4. Torrens, F., & Castellano, G., (2020). Mass spectrometry: The only method with many ionization techniques. In: Pourhashemi, A., Deka, S. C., & Haghi, A. K., (eds.), *Research Methods and Applications in Chemical and Biological Engineering* (pp. 47–57). Apple Academic–CRC Press: Waretown, NJ.
5. Torrens, F., & Castellano, G., (2020). Metaphors that made history: Reflections on philosophy/science/DNA. In: Aguilar, C. N., Ameta, S. C., & Haghi, A. K., (eds.), *Green Chemistry and Biodiversity: Principles, Techniques, and Correlations* (pp. 267–279). Apple Academic–CRC Press: Waretown, NJ.
6. Torrens, F., & Castellano, G., (2007). Book of abstracts. *Gender Bias and Inequality in Evaluating Academic Quality*. Bellaterra, Barcelone, Spain, Universitat Autònoma de Barcelona: Bellaterra, Barcelone, Spain. P-1.
7. Torrens, F., & Castellano, G. (2020). Research with a gender perspective, also in experimental science. In: Esteso, M. A., Ribeiro, A. C. F., & Haghi, A. K., (eds.), *Theoretical and Empirical Analysis in Physical Chemistry: A Framework for Research.* Apple Academic–CRC Press: Waretown, NJ.
8. Torrens, F., & Castellano, G. (2020). Service-learning and employability in physical chemical education, M. A., Ribeiro, A. C. F., & Haghi, A. K., (eds.), *Theoretical and Empirical Analysis in Physical Chemistry: A Framework for Research.* Apple Academic–CRC Press: Waretown, NJ.
9. Torrens, F., & Castellano, C. (2020). Computation, quantum mechanics, algorithms, simulations and tools, M. A., Ribeiro, A. C. F., & Haghi, A. K., (eds.), *Theoretical and Empirical Analysis in Physical Chemistry: A Framework for Research.* Apple Academic–CRC Press: Waretown, NJ.
10. Torrens, F., & Castellano, G. (2020). MOFs, science, pseudoscience, and chronobiology. In: Kulkarni, S., Rawat, N. K., & Haghi, A. K., (eds.), *Green Chemistry and Green Engineering: Processing, Technologies, Properties, and Applications*. Apple Academic–CRC Press: Waretown, NJ.

11. May, P., & Cotton, S., (2015). *Molecules That Amaze Us*. CRC Press: Boca Raton. FL.
12. Fuentes, D., (2016). Aging theories: Free radicals and antioxidants. *Daualdeu, 2016*(10), 14–15.
13. Tortajada, G. L. A., (2019). *Personal Communication.*
14. Botella, V., (2016). The women of the future. *Daualdeu, 2016*(10), 9–10.
15. Rueda, P. S., (2019). Book of Abstracts, Contributions and Difficulties of Women and LGTB+ People in the Science, Sao-Burjassot, Burjassot, València, Spain, October 14, , Saó-Burjassot, 2019, Burjassot, València, Spain, 2019; O-1.
16. Furió, M. C., & Furió, G. C., (2016). Algunes reflexions docents sobre una qüestió controvertida: La motivació dels estudiants envers la química, *Daualdeu*, 2016(10), 19–21.
17. Gosálvez, I., (2019). *Personal Communication.*

CHAPTER 13

Women, Men, Health, Big Data, Taste to Classify and Innovation

FRANCISCO TORRENS[1] and GLORIA CASTELLANO[2]

[1]*Institut Universitari de Ciència Molecular, Universitat de València, Edifici d'Instituts de Paterna, P.O. Box 22085, E-46071 València, Spain, E-mail: torrens@uv.es*

[2]*Departamento de Ciencias Experimentales y Matemáticas, Facultad de Veterinaria y Ciencias Experimentales, Universidad Católica de Valencia San Vicente Mártir, Guillem de Castro-94, E-46001 València, Spain*

ABSTRACT

Big data are data parallelism, not model parallelism. The sector that is investing the most in big data analytics in banking. Data have become an asset: They are key to the success of digital organizations. Machine learning is to computation what probability and statistics are to mathematics. Machine learning does not do magic: Without good data, it will not work and, the more data, the better; but data are not everything. Cloud computing allows combining measurement data from multiple remote devices, and the deployment of machine learning models without modifying the device or software at the edge. Work hard, work smart: Distinguish the important from the urgent. Do it with passion or not at all. Success is 99% failure. Equilibrium: Perseverance ↔ Flexibility. Once you stop learning, you start dying. Desire to be *different*. Be brave even if you are not: You fail until you get it. (Be) live in (you)self. The importance of networking. You do not get extra credit for doing it alone. Never underestimate the power of a mentor. The hard part is managing diversity. Society does not forgive errors but they are the engine that moves us forward.

13.1 INTRODUCTION

Setting the scene: bioinformatics (BI), why the computer is also a laboratory, gender differences in health, big data (BD), BD analytics (BDA), artificial intelligence (AI) making sense of taste [electronic tongues (ETs)] to classify liquids, innovation, a personal view, towards an AI by and for Society. The BD is important for women, men, and health (*cf.* Figure 13.1).

FIGURE 13.1 Big data.

Dynamical Systems and Computational Virology (DySCoVir) Associated Unit Institute for Integrative Systems Biology–Centre de Recerca Matemàtica (I^2SysBio-CRM) celebrated *Kick-off Meeting* on October 4, 2019 (*cf.* Figure 13.2).

FIGURE 13.2 Dynamical systems and computational virology I^2SysBio–CRM.

In earlier publications, it was informed the empirical didactics of molecular shape [1], the phylogenesis of anthropoid apes [2], the fractal analysis of proteins' tertiary structure (P3S) [3], fractal hybrid orbitals (FHOs) in biopolymer chains [4], FHOs in protein models [5], FHOs P3S analysis [6], resonance in interacting induced-dipole polarizing force fields (FFs), application to FF derivatives [7], the modeling of complex multicellular systems, tumour-immune cells competition [8], molecular diversity classification *via* information theory [9], a tool for the interrogation of macromolecular structure [10], a tool for the study of resonance in chemical education [11], dialectic walk on science [12], Brownian motion, random trajectory, diffusion, fractals, chaos theory, dialectics [13], the work with nanomaterials, reductionism, positivism philosophical, ethical considerations [14], BD in AI, models for health, healthcare [15], disruptive BD, AI, physics and process understanding [16]. The purpose of the present report is to review BI, why the computer is also a laboratory, gender differences in health, BD, BDA, AI making sense of taste to classify liquids, innovation, a personal view, towards an AI by and for Society. The aim of this work is to initiate a debate by suggesting a number of questions (Q), which can arise when addressing subjects of why the computer is also a laboratory, gender differences in health and BD, and providing, when possible, answers (A) and hypotheses (H).

13.2 BIOINFORMATICS, OR WHY THE COMPUTER IS ALSO A LABORATORY

Arzalluz proposed questions/answers on BI and why the computer is a laboratory (*cf.* Figure 13.3) [17].

Q1. Why is the computer also a laboratory?
Q2. Where do we come from?
Q3. Where do we go?
Q4. Where can BD be stored?
Q5. What information can be extracted from BD?
Q6. How can information be extracted from BD?
Q7. Have people professionals able to extract information from BD?
Q8. What are biological BD?
Q9. Where do biological BD come from?
Q10. Especially, where do biological BD go?
A10. BI.

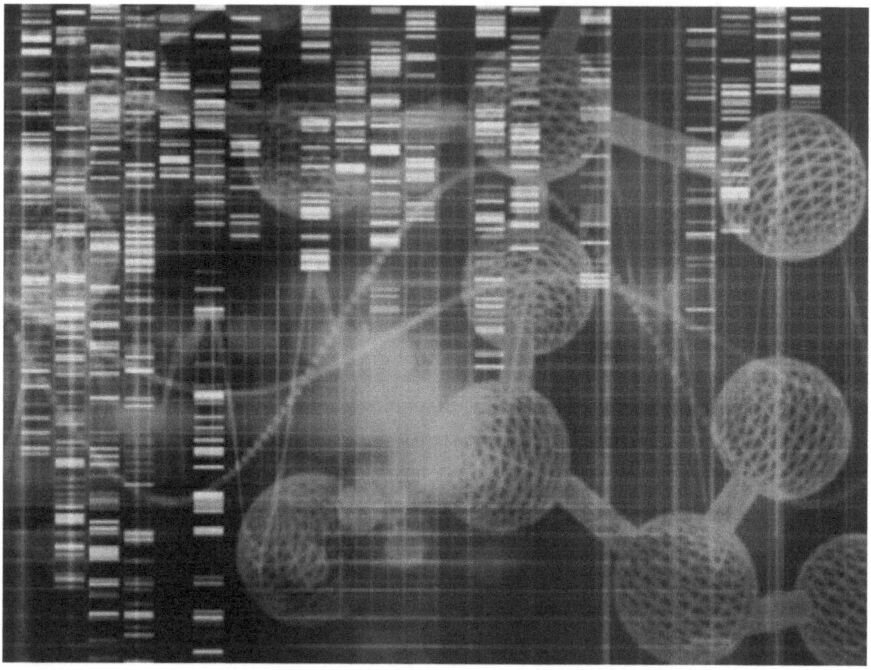

FIGURE 13.3 Bioinformatics, or why the computer is also a laboratory.

13.3 GENDER DIFFERENCES IN HEALTH AND BIG DATA

García proposed some questions and answer on gender differences in health and BD (*cf.* Figure 13.4) [18].

Q1. Why are there some diseases that are more prevalent or serious in men or women?
Q2. Do loneliness and health level present the same pattern by gender?
A2. Processing and analysis of BD will allow more precision, personalization, and optimization.

13.4 BIG DATA ANALYTICS

Gómez Adrián organized a day on BDA. The following hypotheses were proposed [19].

H1. BD is data parallelism, not model parallelism.

H2. The sector that is investing the most in BDA is banking.
H3. Data have become an asset: They are the key to the success of digital organizations.
H4. Machine learning (ML) is to computation what probability and statistics are to mathematics.
H5. ML does not do magic: Without good data, it will not work and, the more data, the better; but data are not everything.

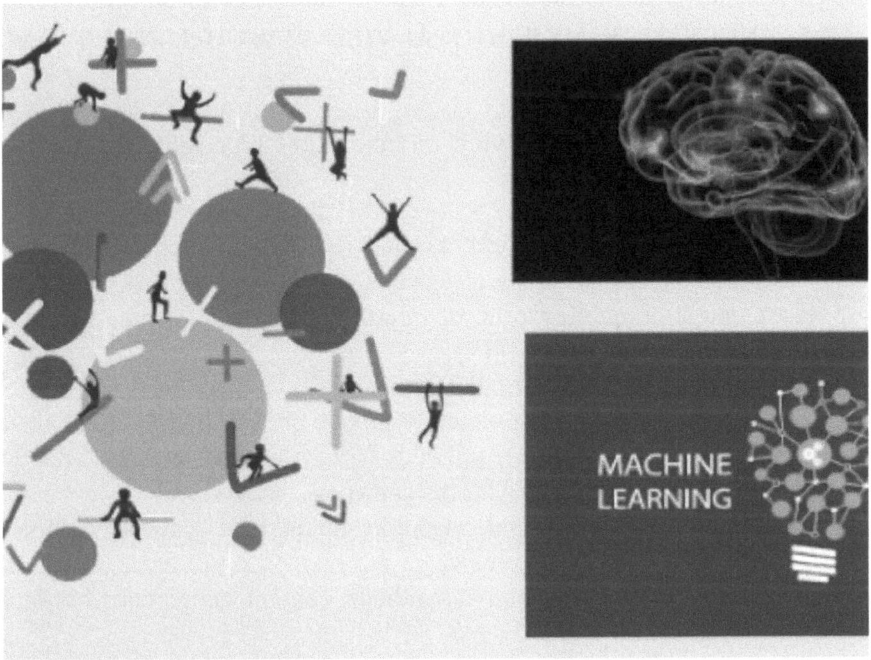

FIGURE 13.4 Gender differences in health and big data.

13.5 ARTIFICIAL INTELLIGENCE MAKES SENSE OF TASTE TO CLASSIFY LIQUIDS

The ETs based on potentiometry offers the prospect of rapid and continuous chemical fingerprinting for portable and remote systems. Ruch group informed a technology platform, including a miniaturized ET based on electropolymerized ion-sensitive films, microcontroller-based data acquisition, a smartphone interface, and cloud computing back-end for data storage and deployment of ML models [20]. The sensor array recorded a series of differential voltages without the use of a true reference electrode, and the resulting time-series potentiometry data was used to train supervised ML algorithms. For trained systems, inferencing tasks, for example, classification of liquids, were realized within less than 1m including data acquisition at the edge and inference *via* the cloud-deployed ML model. They reported the preliminary demonstration of the complete ET technology stack for the classification of beverages and mineral water.

13.6 INNOVATION: MY PERSONAL VIEW TOWARDS AN AI BY AND FOR SOCIETY

Oliver presented her personal view on innovation, proposing the following hypotheses on AI [21].

H1. Challenges: aging world; limited resources; which are unevenly allocated; climate crisis.
H2. Context: a mobile world; of Internet-of-things; of social media; 4th Industrial Revolution (4IR).
H3. 4IR: AI is at the core of 4IR.
H4. Two schools of AI: top-down, symbolic or neat; bottom-up, connectionist, or scruffy.
H5. Specific properties of AI: pros and cons.
H6. Pros of AI: transversal, invisible, ubiquitous; scalable, complex, updateable; ability to predict.
H7. Cons of AI: asymmetry; synthetic content generation; hackable (*adversarial* ML).
H8. Data explosion: People live in a world of data; computing power; learning models.
H9: Area: pattern recognition from unstructured data.
H10. AI has become a political issue.
H11. AI is part of the national and supranational strategies.
H12. Five key dimensions: technology; social; ethics; human; economy and labor.
H13. Equation: Innovation = Invention + Communication.
She proposed the following question and hypotheses on what she has learned.
Q1. What have I learned?
H14. Work hard, work smart: Distinguish the important from the urgent.
H15. Do it with passion or not at all.

H16. (Soichiro Honda). Success is 99% failure.
H17. Perseverance. Equilibrium: Perseverance ↔ Flexibility.
H18. (Albert Einstein). Once you stop learning, you start dying.
H19. Desire to be *different*.
H20. Be brave even if you are not: You fail until you get it.
H21. (Be)live in (you)self.
H22. The importance of networking.
H23. (Linda Austin). You do not get extra credit for doing it alone.
H24. Never underestimate the power of a mentor.
H25. Luck is what happens when preparation meets opportunity.
H26. Code of ethics.
H27. My FATEN project.
H28. FATEN–F: *Fairness* or justice, non-discrimination, cooperation.
H29. FATEN–A: Autonomy, accountability, intelligent augmentation.
H30. FATEN–T: Trust and Transparency.
H31. FATEN–E: Eneficence; progress, sustainability, diversity, veracity; education.
H32. FATEN–N: Non-maleficence: reliability, security, reproducibility.
H33. Have fun!
H34. (Reis Juan). In a company, the hard part is managing diversity.
H35. (Reis Juan). Society does not forgive errors but they are the engine that moves us forward.

13.7 FINAL REMARKS

From the present results and discussion, the following final remarks can be drawn.

1. Big data are data parallelism, not model parallelism. The sector that is investing the most in big data analytics in banking. Data have become an asset: They are key to the success of digital organizations. Machine learning is to computation what probability and statistics are to mathematics. Machine learning does not do magic: Without good data, it will not work and, the more data, the better; but data are not everything.
2. The incorporation of cloud computing services allows combining measurement data from multiple remote devices, and the deployment of machine learning models without modifying the device or software at the edge.
3. Work hard, work smart: Distinguish the important from the urgent.

4. Do it with passion or not at all.
5. Success is 99 percent failure.
6. Perseverance. Equilibrium: Perseverance ↔ Flexibility.
7. Once you stop learning, you start dying.
8. Desire to be different.
9. Be brave even if you are not: You fail until you get it.
10. (Be)live in (you)self.
11. The importance of networking.
12. You do not get extra credit for doing it alone.
13. Never underestimate the power of a mentor.
14. Luck is what happens when preparation meets opportunity.
15. Code of ethics. Oliver's FATEN project. Have fun!
16. In a company, the hard part is managing diversity. Society does not forgive errors but they are the engine that moves us forward.

ACKNOWLEDGMENTS

The authors thanks the support from Fundacion Universidad Catolica de Valencia San Vicente Martir (Project No. 2019-217-001UCV).

KEYWORDS

- **bioinformatics**
- **cloud computing**
- **computer**
- **electrodeposition**
- **electronic tongue**
- **polymer**
- **smartphone**

REFERENCES

1. Torrens, F., Sánchez-Pérez, E., & Sánchez-Marín, J., (1989). Empirical teaching of the molecular form. Science Teaching, Extra Number (III Congress), (1), 267-268.
2. Torrens, F., (2000). Phylogenesis of anthropoid apes. Encounters in Biology, *8*(60), 3–5.

3. Torrens, F., (2000). Fractal analysis of the tertiary structure of proteins.Encounters in Biology, *8*(64), 4–6.
4. Torrens, F., (2000). Fractal hybrid orbitals in biopolymer chains. *Russ. J. Phys. Chem. (Engl. Trans* L.), *74*, 115–120.
5. Torrens, F., (2001). Fractals for hybrid orbital's in protein models. *Complexity Int., 8*, torren01–1–13.
6. Torrens, F., (2002). Fractal hybrid orbitals analysis of the tertiary structure of protein molecules. *Molecules, 7*, 26–37.
7. Torrens, F., & Castellano, G., (2009). Resonance in interacting induced-dipole polarizing force fields: Application to force-field derivatives. *Algorithms, 2*, 437–447.
8. Torrens, F., & Castellano, G., (2009). Modeling of complex multi-cellular systems: Tumor-immune cell competition. *Chem. Central J., 3*(Suppl. I), 75–1–1.
9. Torrens, F., & Castellano, G., (2012). Molecular diversity classification *via* information theory: A review. *ICST Trans. Complex Syst., 12*(10–12), e4–1–8.
10. Torrens, F., & Castellano, G., (2014). A tool for interrogation of macromolecular structure. *J. Mater. Sci. Eng. B, 4*(2), 55–63.
11. Torrens, F., & Castellano, G., (2014). A new tool for the study of resonance in chemical teaching. *Advances in Science and Engineering, 5*(1), 81–91.
12. Torrens, F., & Castellano, G., (2017). Dialectic walk on science. In: Laguarda-Miro, N., Masot-Peris, R., & Brun-Sánchez, E., (eds.), *Sensors and Molecular Recognition* (Vol. 11, pp. 271–275). Universidad Politécnica de Valencia: València, Spain.
13. Torrens, F., & Castellano, G., (2019). Brownian motion, random trajectory, diffusion, fractals, theory of chaos, and dialectics. In: Haghi, R., Besalú, E., Jaroszewski, M., Thomas, S., & Praveen, K. M., (eds.), *Modern Physical Chemistry: Engineering Models, Materials, and Methods with Applications* (pp. 3–12). Apple Academic–CRC Press: Waretown, NJ.
14. Torrens, F., & Castellano, G., (2018). Working with nanomaterials. Philosophical considerations: Reductionism/positivism and ethics. In: Feltrero, R., (ed.), *Technology and Social Innovation: Towards an Inclusive and Sustainable Development* (pp. 11–35). Intellectual Challenges of the XXI Century No. 1, Global Knowledge Academics: Cantoblanco, Madrid, Spain.
15. Torrens, F., & Castellano, G., (2020). Big data in artificial knowledge: Models for health and healthcare. In: Torrens, F., Mahapatra, D. K., & Haghi, A. K., (eds.), *Physical Biochemistry, Biophysics, and Molecular Chemistry: Applied Research and Interactions* (pp. 69–82). Apple Academic–CRC Press: Waretown, NJ.
16. Torrens, F., & Castellano, G., (2021). Disruptive big data, intelligence and physics: Process understanding. In: Khalaf, M. N., Smirnov, M. O., Kannan, P., & Haghi, A. K., (eds.), *Environmental Technology and Engineering Techniques: Basic Concepts and Health Interventions* (pp. 167–174). Apple Academic–CRC Press: Waretown, NJ.
17. Arzalluz, Á., (2019). *Personal Communication.*
18. García, F., (2019). *Personal Communication.*
19. Gómez Adrián, J. A., (2019). Book of Abstracts, Jornada Big Data Analytics, València, Spain, September 19, 2019; Universidad Politécnica de Valencia: València, Spain.
20. Ruch, P. W., Hu, R., Capua, L., Temiz, Y., Paredes, S., Lopez, M. A., Barroso, C. J., et al., (2019). A portable potentiometric electronic tongue leveraging smartphone and cloud platforms. *arXiv, 2019*, 1907.06641–1–3.
21. Oliver, N., (2019). *Personal Communication.*

Index

B